Quantum Mechanics

PHYSICS AND ITS APPLICATIONS

Series Editors

E.R. Dobbs
University of London

S.B. Palmer
Warwick University

This series of short texts on advanced topics for students, scientists and engineers will appeal to readers seeking to broaden their knowledge of the physics underlying modern technology.

Each text provides a concise review of the fundamental physics and current developments in the area, with references to treatises and the primary literature to facilitate further study, Additionally texts providing a core course in physics are included to form a ready reference collection.

The rapid pace of technological change today is based on the most recent scientific advances. This series is, therefore, particularly suitable for those engaged in research and development, who frequently require a rapid summary of another topic in physics or a new application of physical principles in their work. Many of the texts will also be suitable for final year undergraduate and postgraduate courses.

Quantum Mechanics

Second edition

Paul C.W. Davies

Professor of Natural Philosophy
University of Adelaide
Australia

David S. Betts

Honorary Reader in Experimental Physics
University of Sussex
Brighton, UK

First edition published by Chapman & Hall 1984
Second edition published by Chapman & Hall 1994
Reprinted in 1999 by Stanley Thornes (Publishers) Ltd

Reprinted in 2002 by:
Nelson Thornes Ltd
Delta Place
27 Bath Road
CHELTENHAM
GL53 7TH
United Kingdom

03 04 05 / 10 9 8 7 6 5 4 3

A catalogue record for this book is available from the British Library

ISBN 0 7487 4446 0

Page make-up by Thomson Press (India) Ltd, New Delhi

Printed and bound in Spain by GraphyCems

Contents

Preface

What is quantum mechanics? The laws of Newton, which work well over a wide range of scale, from planets to billiard balls, fail at the level of atoms and their constituents. Quantum mechanics provides a new set of laws and a mode of description for microscopic systems that has proved supremely successful. Since its inception in the 1920s it has explained the structure of atoms and molecules, nuclear reactions, collisions of particles in accelerators, the emission and absorption of radiation, the thermal and electrical properties of solids, chemical bonding, superconductivity and superfluidity, the pressure inside collapsed stars, subnuclear matter, and a great deal else.

Why is quantum mechanics so important? Besides providing explanations for a wide variety of physical phenomena, the theory has led to a large number of practical applications, among which are the laser, the electron microscope and the microchips of modern electronics. In many research laboratories and high technology industries, devices based on quantum principles are used or developed as a matter of routine. Although the foundations of the subject are well over half a century old, new applications of the theory still have the power to surprise.

Why is it so hard to learn? Students find quantum mechanics tough going for two reasons, one conceptual, the other technical. Familiar concepts such as speed, size, acceleration, momentum and energy take on weird features, or even become meaningless. Intuition gained from daily experience is of no help, and can be misleading. The student must learn to think about mechanical concepts in a completely different way. Some of these conceptual issues are still a matter of dispute among physicists and have raised fundamental philosophical questions about the nature of reality and the role of the observer in the physical universe. Readers who wish to read an elementary account of these topics might turn to A. I. M. Rae's *Quantum Physics: Illusion or Reality?* (Cambridge University Press, 1986) or J. S. Bell's *Speakable and Unspeakable in Quantum Mechanics* (Cambridge University Press, 1987).

On the technical side, the mathematical description of quantum processes is rather abstract and not very obviously related to the subject of its description. Physical quantities are represented by mathematical objects with unusual properties. Some of the mathematics is also often new to the student and learning it can be an additional burden.

In this book we give all the essentials and some of the applications. The treatment is suitable for science undergraduates with no previous knowledge of the quantum theory and takes them up to a typical degree level. It is intended that this single book should provide all that most physics undergraduates will need.

The first edition was written by one of us (Paul Davies) in 1984, and the ideas which generated its characteristic style and choice of topics were outlined in its preface. Those ideas are shared by the second author (David Betts), and readers familiar with the first edition will find that almost nothing of the original has been removed. There are some new chapters, however, notably on areas which are important in applications. Arrays of potential wells are crucial to an understanding of some properties of solids, for example semiconduction, and for this reason we have devoted Chapter 3 to considerations of one-dimensional arrays of simple wells. Scattering by steps and barriers in one dimension was included in the first edition, but we have sought to make this more interesting and relevant by the addition of Chapter 5 on scattering in three dimensions. Atomic physics is traditionally a key application of quantum mechanics and is now included here as Chapter 9. These chapters are embedded in the earlier material which has been altered and expanded in a small number of places. There have also been substantial enhancements to the sequences of figures and exercises. The result is a noticeably enlarged volume, although we have tried to retain the direct, terse style of the original, believing that to be a distinct merit.

Although we have deliberately adopted a terse style, our intention is to go beyond a mere catalogue of facts and procedures to provide an understanding of the principles underlying quantum mechanics. Readers who get discouraged should soldier on. Quantum mechanics is a subject that usually comes right in the end, even though it can seem horribly obscure when only half-learned.

We hope that users of the first edition will feel pleased with this new version, and that new users will come to value its combination of enthusiasm, style, coverage and price.

ACKNOWLEDGEMENTS

We are grateful to Professor Roland Dobbs for his help in suggesting improvements both before work started and in the final stages.

1

Preliminary concepts

This chapter presents the basic physical ideas that underpin the quantum theory. The central features are uncertainty and discreteness. The conceptual foundations of the theory involve some notorious subtleties, but these need not get in the way of progress. The reader may find it helpful to return to this chapter having mastered the later technical sections.

1.1 ORIGINS

Historians place the beginnings of the quantum theory at 1900. That year Max Planck published his famous formula for the distribution of energy as a function of frequency in black-body radiation. The details need not concern us. It was here that a new fundamental constant was introduced into physics – Planck's constant h. More often we shall use $h/2\pi$, denoted by $\hbar$.

Although Planck did not fully grasp the significance of this step, with the benefit of hindsight the pertinent features are as follows. Electromagnetic energy, which prior to these developments was regarded as continuous, in fact comes in discrete lumps or packets – quanta – called, in this case, photons. The discrete nature of radiation manifests itself in the processes of absorption or emission by matter, which occurs only in multiple units of these energy packets. The energy is not the same for all photons but is related to the frequency v of the electromagnetic wave by

$$E = hv \qquad (1.1)$$

or, equivalently, $E = \hbar\omega$, where $\omega = 2\pi v$, a fundamental formula of quantum theory. Note carefully that the high-frequency (short-wavelength) photons are the most energetic. For light, $v \approx 10^{15}$ Hz, so $E = 10^{-18}$ J – negligible in the macroscopic world. Only at the atomic level is equation (1.1) of importance.

A further important formula follows from the relation between energy E, momentum p, and rest mass m according to the theory of relativity:

$$E^2 = p^2c^2 + m^2c^4 \tag{1.2}$$

Classical electromagnetic theory connects E and p for an electromagnetic wave through the relation

$$E = pc \tag{1.3}$$

where $p = |p|$. Comparing equations (1.2) and (1.3) reveals that for a photon $m = 0$. For this reason the photon is sometimes called a 'massless particle'. From equations (1.1) and (1.3) we then arrive at

$$p = \frac{hv}{c} = \frac{h}{\lambda} = \hbar k \tag{1.4}$$

where λ is the wavelength and k the wave number.

Students sometimes ask why electromagnetic energy should manifest itself in packets. It is important to realize that this phenomenon cannot in some way be derived from classical electromagnetic theory. It must be accepted as a simple fact. That is the way nature is.

Curiously, although radiation theory gave birth to quantum mechanics, the interaction of matter and electromagnetic radiation is a rather advanced topic in quantum mechanics and it took almost fifty years for an acceptable formulation to be achieved. One reason for this is that photons are quanta of the electromagnetic field, and the quantum theory of fields requires a relativistic treatment to be entirely successful.

Confirmation of the photon hypothesis comes from the photoelectric effect. A stream of radiation, when impinging on a metal surface, can knock electrons out after the fashion of a coconut shy. Generally, one photon knocks out one electron. We see this by increasing the flux; more electrons come out because more photons are arriving. Similarly, increasing the frequency, the electrons emerge with greater energy, in accordance with equation (1.1). These features would be perplexing on the basis of the old classical theory of electromagnetism.

1.2 COLLAPSE OF DETERMINISM

The discrete character of electromagnetic energy leads directly to the need for a profound modification in our understanding of the physical world. Consider a *polarized* light wave which encounters a polarizer astride its direction of propagation. If the polarizer is oriented parallel to the direction of polarization, all the light is transmitted; if the polarizer is perpendicular to the direction of polarization, none is transmitted. At 45° half the light gets through. Generally the transmitted fraction is $\cos^2 \theta$, where θ is the angle of the polarizer axis relative to the polarization plane (Fig. 1.1).

All this is readily comprehensible according to Maxwell's electromagnetic theory, using vector projections; at 45° the polarizer takes out half the wave energy. Weird overtones develop, however, when the quantum nature of light is taken into account. Suppose the intensity of the light is reduced so that only one photon at a time arrives at the polarizer. Because photons cannot be chopped in half, each photon either does, or does not, get through the polarizer. That much is obvious. But there is nothing to distinguish any one photon from any other – the polarizer has no means of sorting them into 'sheep' and 'goats'. If it sometimes passes a photon and sometimes blocks it, without rhyme or reason, all we can say is that there is a fifty-fifty chance of any given photon being passed. Thus, on average, half the photons get through and half do not. We may say that a particular photon will be transmitted through the polarizer with a *probability* of 0.5. As the orientation θ is varied, so the odds in favour of transmission vary as $\cos^2 \theta$.

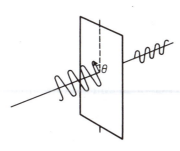

Fig. 1.1 When plane-polarized light encounters an obliquely oriented polarizer only a fraction $\cos^2 \theta$ of the intensity is transmitted, corresponding to that component of the polarization vector that lies along the polarization axis. In quantum language this implies that only a fraction of the incident photons is transmitted, each photon with probability $\cos^2 \theta$.

The injection of this fundamental probabilistic element into physics is a major conceptual step. Prior to quantum theory, physics was ultimately *deterministic*. The behaviour of every physical system was supposed, in principle, to be determined in advance. According to Maxwell's classical electromagnetic theory, a given light wave pattern encountering a given polarizing material would inevitably and unalterably produce the same result. The quantum factor, by contrast, implies that we can never know in advance what is going to happen or how the light will develop. Only the betting odds can be given. We shall see that this indeterminism is a universal feature of the micro-world.

1.3 MATTER WAVES

The corpuscular nature of light was only half of the input that gave birth to quantum mechanics. The other half concerns matter. Given that the discreteness of photons manifests itself during the emission and absorption process, it is not surprising to find that quantum oddities afflict the behaviour of material particles too. That something strange happens to atomic particles was already clear from severe difficulties encountered with Rutherford's planetary model of the atom. Because the electrons orbiting the nucleus are accelerating, they should, according to classical electrodynamics, continually emit electromagnetic radiation and rapidly spiral into the nucleus. All atoms should be unstable and collapse amid a shower of radiation, i.e. classical electrodynamics fails to explain the stability of atoms. Furthermore, it cannot account for the *discrete frequencies* of radiation emitted by atoms which give rise to their characteristic spectral lines. According to the classical model the cavorting electrons ought to generate a whole range of electromagnetic wave frequencies corresponding to their unrestricted orbital frequencies. In practice, though, a typical spectrum features sequences of sharp lines arranged in an orderly fashion, indicating a complicated but not haphazard internal organization of the atom.

In 1912 Niels Bohr used Planck's idea of quantum discreteness to produce a model of the atom in which the bound electrons are inexplicably restricted to certain discrete *energy levels*. Transitions between levels can occur, but only with the emission or absorption of a photon. For example, in the hydrogen atom the positioning of the energy levels is given by the simple formula

$$E_n \propto -1/n^2 \tag{1.5}$$

where $n = 1, 2, 3. \ldots$. The frequency of the photon emitted when the electron makes a transition from level n to level m is then restricted to the discrete values

$$h\nu = (E_n - E_m) \propto (m^{-2} - n^{-2}) \tag{1.6}$$

a formula in close accord with the evidence of spectroscopy.

Though Bohr's so-called 'old' quantum theory was undoubtedly on the right track, it made no attempt to explain the *mechanical* principles which compel the atomic electrons to reside in the designated energy levels, nor did it explain how or why they can jump between the levels. It was not until the mid-1920s that a proper *quantum mechanics* was developed. The clue to this new mechanics came from the realization that particles such as electrons have certain characteristics reminiscent of *waves*.

With the benefit of hindsight we can now see that the experimental evidence for the wave-like nature of matter had been present for some time in the work of Clinton Davisson. For many years Davisson had experimented with the scattering of electron beams from crystal surfaces. A crystal lattice forms a regular array from which the electrons rebound. Instead of scattering randomly, the electrons come out in an organized pattern. Particularly after his 1927 experiments, it became clear that these patterns were exactly like wave diffraction.

Already in 1924 Louis de Broglie had suggested that atomic particles might have a wave-like aspect to their behaviour. He linked the momentum of the particle to the wavelength of those 'matter waves' by

$$p = h/\lambda \tag{1.7}$$

as for light (equation (1.4)). Thus Planck's constant and the quantum are introduced into the behaviour of matter too.

Although the early experiments were performed with electrons, the wave-like association applies to all particles. De Broglie waves exist for protons, neutrons, mesons and even whole atoms and molecules. In fact there is no scale of size at which the wave-like property abruptly disappears, although relation (1.7) shows that as the mass (hence p) becomes large, λ becomes very small, and the length scale over which wave-like behaviour can be discerned shrinks correspondingly. Thus at room tempeature an average atom of oxygen has a de Broglie wavelength of only 4×10^{-11} m. For a molecule of DNA $\lambda \approx 10^{-14}$ m. Obviously for macroscopic objects

λ is so small as to render the wave-like aspects of matter utterly negligible. That is why we do not notice matter waves in daily life.

1.4 HEISENBERG'S UNCERTAINTY PRINCIPLE

There seems to be a conceptual conflict between the notion of, say, an electron or an atom as a microscopic particle and the notion of a wave. One clue to reconciling these two disparate aspects of description comes from a careful study of just how one might go about establishing the particle-like nature of an electron.

What is a particle? Two characteristics seem paramount. At all times a particle must be at a place, i.e. have a definite location, and it must have a motion, i.e. be going somewhere, or be at rest. Putting these two together, our minimal requirement for the designation 'particle' is that the electron or atom should follow a well-defined trajectory in space and time. A variety of thought experiments were devised to analyse how one might in principle observe these two crucial features.

Suppose, for example, one wishes to measure the speed v of the particle. One way of doing this is to examine the frequency of electromagnetic radiation bounced off the particle (such as police radar speed-traps for cars) or, better still, emitted by the particle, and then to use the Doppler shift. If an atom emits light of frequency v_0 in its rest frame, and it is moving relative to the observer at speed v, then to first order in v/c the received frequency is Doppler-shifted to

$$v \approx v_0(1 + v/c) \tag{1.8}$$

Therefore, a measurement of v is a measurement of v. However to measure v it is necessary for many cycles of the wave to occur. For a measurement of finite duration τ one knows from wave theory that there is an unavoidable uncertainty or error in the frequency of at least

$$\Delta v \approx 1/\tau \tag{1.9}$$

This would not be a serious limitation if τ could be made arbitrarily long, but it is at this stage that the quantum nature of light intervenes. The atom will suffer a disturbance as a result of the radiation emission and this will cause the atom to change its trajectory from what it would otherwise have been in the absence of disturbance. This change could be taken into account if we knew exactly how and when the disturbance occurred. The trouble is that because of the discrete nature of the photons the emission of radiation occurs all in one go, as a single photon, and the atom recoils abruptly as a consequence. However, we have no idea at what moment during the interval τ this recoil event occurs. For a photon of momentum $h\nu/c$ the recoil speed of a particle of mass m will be $v = h\nu/mc$, and in a time τ the position of the particle could alter from its original trajectory by as much as $h\nu\tau/mc$. There is thus a fundamental uncertainty in the position of the particle

$$\Delta x \approx h\nu\tau/mc = h\nu/\Delta v\,mc \qquad (1.10)$$

using equation (1.9). From equation (1.8) we see that an error Δv in v amounts to an error $v_0 \Delta v/c \approx v\Delta v/c$ (as $v \approx v_0$ for small v/c). Thus equation (1.10) may be rearranged to yield

$$\Delta x\, \Delta p \approx h \qquad (1.11)$$

where $p = mv$ is the momentum of the particle. This is Werner Heisenberg's famous uncertainty relation. It states that one cannot simultaneously measure position and momentum to arbitrary precision. Better knowledge of one must be traded for ignorance of the other. Crude derivations of this sort cannot yield a precise numerical factor in equation (1.11), which is why we use the $\approx$ sign. For that reason one often sees $\hbar$ rather than h on the right-hand side. More refined arguments can yield a definite numerical factor.

It must not be supposed, in view of this particular derivation of equation (1.11), that the quantum uncertainty is somehow purely the result of an attempt to effect a measurement – a sort of unavoidable clumsiness in probing delicate systems. The uncertainty is *inherent* in the microsystem – it is there all the time, whether or not we actually choose to measure x or p.

In fact the intrinsic uncertainty in the behaviour of a particle is built into the wave-like character of its propagation. If we wish to be sure a particle is confined to a region Δx then we must suppose its associated de Broglie wave is squashed into this region. Such a packet cannot, however, be built out of waves of just one wavelength. In fact, Fourier analysis shows that a wave packet of size Δx contains a spread of wave numbers $\Delta k \approx 1/\Delta x$. Recalling equation (1.7), one recovers equation (1.11).

One consequence of the uncertainty principle is that a photon specified by a precise value of its frequency v has a precise momentum $p = h\nu/c$, so its position is completely indeterminate – it is not localized in space at all. Another important result is that a particle such as an electron cannot have a well-defined path through spacetime. The concept of a trajectory implies that, at any given instant, one may specify where a particle is and the direction and speed with which it is moving. As we have seen, these are incompatible requirements in quantum mechanics. We may know that at one time the electron is at place A and at a later time it is at B, but we cannot say exactly how it got from A to B; we cannot trace out a well-defined path of motion connecting the two points. Indeed, there is a sense in which the particle takes *all possible paths* together.

A clear illustration of the abstract nature of particle waves is given by the following thought experiment. Imagine a particle confined somewhere in a box; we do not know exactly where, so the wave must be spread throughout the box. An impenetrable membrane is now inserted, dividing the box into two disconnected chambers A and B. Some of the wave is trapped in A, some in B. The particle, however, is obviously *either* in A or in B, even though the wave is in *both*. If an observation is now made and the particle is seen in, say, chamber A, then the wave in chamber B must abruptly disappear because there is now zero probability of the particle residing in B. The behaviour of the wave, while undeniably subject to physical principles (it has an equation of motion, at least when observations are not being made), also in some sense reflects our *knowledge* of the system.

The discussion given in this section thus associates the wavelike aspect of a particle with the irreducible fuzziness in its particle-like properties. It is as though the wave represents a sort of area of ignorance surrounding the particle. This suggests that we extend the probabilistic character of photons to electrons and other material particles, and that the matter waves somehow encode the statistical nature of this probability. We cannot know, on the basis of the analysis so far, exactly how this encoding is achieved. That step requires a specific hypothesis which can then be tested experimentally. We shall see which hypothesis was successful when we get to section 1.6, but first some clarification of the relationship between the wave- and particle like aspects of nature must be given.

1.5 WHEN IS A WAVE A PARTICLE?

It will be evident that in some cases, e.g. photoelectric effect, photons may be treated just like particles. Yet we know that light is an electromagnetic *wave.* Similarly electrons and other particles can behave in some respects like waves. Can something be both a wave *and* a particle, or must we make a choice between them?

First it is important to understand what is *not* being claimed. There is no claim that a photon *is* a particle, only that it can sometimes display particle-like behaviour. Similarly, we do not claim an electron *is* a wave, only that it can have wave-like aspects. What does this mean?

As usually visualized, a particle has a well-defined trajectory or world line. We shall see, however, that a photon does not in general possess these properties. It cannot be localized at a place. You can know roughly when a photon hits a target, but that only tells you about the *interaction* between the photon and the target, which can happen at a definite place. Prior to this the photon has no meaningful position.

Turning now to matter waves, resist at all costs the temptation to think of an electron as pulled asunder and smeared out in space in little ripples. The electron itself is not a wave. Rather, the way it moves about is controlled by wave-like principles. Physicists still regard an electron as a point-like entity but the precise location of that point may not be well-defined. What, then, are these matter waves?

They are not waves of any substance but are *abstract* waves. This is not as esoteric as it may at first seem. We are used to talking of crime waves and have no difficulty in understanding them as an abstract concept. Crime waves are not waves of undulating stuff but *probability* waves. Where the wave is most intense, there one has the greatest chance of a felony. Crime waves, like fashions or unemployment, may move about – they have dynamics – but an individual crime still occurs, of course, at a place. It is the abstract probability which moves.

Matter waves are also probability waves. They tell you where the particles are most likely to be found. It is the probability which has the wave-like behaviour while the particles themselves remain as little lumps, albeit elusively secreted in the wave which guides their progress.

Thus the wave- and particle-like aspects of photons and electrons can peacefully coexist. They are not contradictory but *complementary.* Which face of this wave–particle duality is manifested depends on the sort of question that is asked. This is elegantly illustrated by the case of Young's famous two-slit interference experiment.

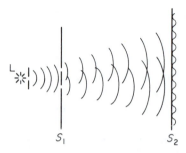

Fig. 1.2 Interference experiment. The source L illuminates neighbouring slits in the screen S, and the image is observed on S_2. Waves diffracting from the apertures overlap alternately in and out of phase at S_2, producing a series of bright and dark fringes, graphed schematically alongside S_2. This experiment may be performed with photons or electrons.

Figure 1.2 shows the conventional set up. A small light source illuminates two closely spaced slits in a screen and the image of the slits is observed on another screen. Light waves from each slit overlap coherently and produce a characteristic pattern of bright and dark interference fringes, bright when the waves arrive at the image screen in phase, dark when they arrive out of phase. The experiment could also be performed with matter waves, e.g. using electrons, and a similar interference pattern would occur.

A careful analysis of the interference pattern shows it to consist of myriads of little spots, each spot produced by the arrival of an electron or photon. In the latter case a permanent record could be kept by replacing the image screen by a photographic plate. If the intensity of the source is turned down so far that only one particle at a time arrives at the plate, then the interference pattern will build up slowly with time in a speckled sort of way as each photon arrives at the screen and makes a dot. Obviously the pattern is a collective or holistic concept – a property of a large collection of dots – and cannot be applied to any individual photon's record. Nevertheless, for the collection to display a pattern each particle must cooperate in a statistical sense. Thus although the final point of arrival of any given particle is indeterminate, there is a greater probability of its going to a bright-fringe region than to a dark-fringe region. It is important to realize that successive particles are completely independent. They produce a pattern not by mutual cooperation but by weighted odds. The pattern would still appear even if one superimposed many different photographic plates from different experiments.

If one of the slits is covered up the interference pattern disappears. The pattern is a product of the whole geometry and requires the superposition of wavelets from both slits. On the other hand, if the electrons or photons are thought of as classical particles, then any given particle can obviously pass through only one of the slits, and the problem is to know how each individual particle can contribute to the interference pattern when it is totally unaware of the existence of *two* slits. It must be repeated that there is no question of an electron coming to bits and spreading about; the particles arrive at the image screen like a hail of shot, each making a mark at one spot. To avoid any conspiracy between different particles, the intensity can be turned down so low that only one particle traverses the apparatus at a time. We must conclude that the *wave* passes through both slits but that the *particle* passes through only one. Because the *behaviour* of the particle is dictated (statistically, somehow) by the wave, the presence of both slits still affects the destination of the particle even though the particle presumably only visits one slit.

Now it might be supposed that if one could determine in each case which slit the particle is heading for, it could not possibly make any difference to its destination if the *other* slit were quickly covered up. However, there is a snag. To find out which slit is being approached one must have a reasonable knowledge of the particle's position. By the Heisenberg uncertainty principle this will introduce a fuzziness in its motion which will smear out the interference pattern. We therefore have two complementary situations. Either we can know the route of the particle through the apparatus, and end up with a featureless image, or we can have an interference pattern but never know which slit any given particle is visiting. When the pattern is there, neither slit can be eliminated.

Should we say that in some sense the particle passes through both slits? A better description is to say that there are two possible worlds, one with the particle going through slit *A*, the other with the particle going through slit *B*. When the pattern is there, *both* these worlds coexist in superposition. There is not merely *either* one reality *or* the other, but an overlapping amalgam of both. This profound and disturbing collapse of the common-sense view of reality in the external world has given rise to decades of debate among physicists and philosophers.

1.6 SCHRÖDINGER'S WAVE EQUATION

If there is a wave there ought to be an equation to go with it. In 1926 Erwin Schrödinger suggested the following equation for a particle of mass m

$$-\frac{\hbar^2}{2m}\nabla^2\psi + V\psi = i\hbar\frac{\partial\psi}{\partial t} \tag{1.12}$$

It is the key equation of the quantum theory and must be accepted as a fundamental postulate. Here, ∇^2 is the Laplacian operator and V is the potential energy, which will generally be a function of both space and time $V(r,t)$. The wave amplitude, which is known as the *wave function*, is likewise a function of r and t, i.e. $\psi = \psi(r,t)$. In one space dimension equation (1.12) reduces to

$$-\frac{\hbar^2}{2m}\frac{\partial^2\psi(x,t)}{\partial x^2} + V(x,t)\psi(x,t) = i\hbar\frac{\partial\psi(x,t)}{\partial t} \tag{1.13}$$

The first problem is to tackle the interpretation of the wave function ψ. This was first accomplished successfully by Max Born. Notice the $i(\equiv\sqrt{-1})$ on the right-hand side of the equation. This is going to make ψ complex. What we observe ought to be real and it is $|\psi|^2$, i.e. $\psi^*\psi$ where * denotes a complex conjugate, that has the physical connection. In accordance with the remarks in the previous sections about probability, $|\psi(r,t)|^2$ is taken to be the *probability density* $P(r,t)$ for a particle to be located at point r at time t. Thus $|\psi(r,t)|^2\,\mathrm{d}\tau$ is the probability it will be in the infinitesimal volume $\mathrm{d}\tau$ at time t. This claim cannot be derived; it is postulated (and works). It provides the precise connection between the fuzziness in our knowledge of a particle and the existence of an associated matter wave which was discussed in an imprecise way in the previous sections. Note carefully that ψ itself is not an observable quantity. For that reason we have some freedom in the form of ψ. Specifically the *phase* of ψ is arbitrary because it cannot be observed; we can always change the phase without changing the observable quantity $|\psi|^2$.

Clearly the *total* probability is 1:

$$\int_{\substack{\text{all}\\\text{space}}} P(r,t)\,\mathrm{d}\tau = \int_{\substack{\text{all}\\\text{space}}} |\psi(r,t)|^2\,\mathrm{d}\tau = 1 \tag{1.14}$$

as the particle must be somewhere. This is called a *normalization* condition. Because every solution of a linear equation such as equation (1.12) can be multiplied by a complex number and remain a solution, equation (1.14) tells us what overall amplitude factor to choose.

Suppose, however, that our integral is over a finite volume $\mathscr{V}$; that gives the probability that the particle is inside $\mathscr{V}$. Because the particle moves about, this probability will change with time – some of the wave might flow in or out of $\mathscr{V}$ carrying the probability with it. To quantify this flow we compute the rate of change of probability for the particle to be in $\mathscr{V}$

$$\frac{\partial}{\partial t} \int_{\mathscr{V}} P \, d\tau = \int_{\mathscr{V}} \left(\psi^* \frac{\partial \psi}{\partial t} + \psi \frac{\partial \psi^*}{\partial t} \right) d\tau \tag{1.15}$$

To keep matters simple we shall resort to a one-dimensional treatment. Using Schrödinger's equation (1.13) and its complex conjugate to eliminate the time derivatives, the right-hand side of equation (1.15) becomes

$$\frac{i\hbar}{2m} \int_{x_1}^{x_2} \left(\psi^* \frac{\partial^2 \psi}{\partial x^2} - \psi \frac{\partial^2 \psi^*}{\partial x^2} \right) dx \tag{1.16}$$

where x_1 and x_2 are the ends of the one-dimensional 'volume' $\mathscr{V}$. Integrating equation (1.16) by parts, only the boundary terms survive. Thus

$$\frac{\partial}{\partial t} \int_{x_1}^{x_2} P \, dx = \frac{i\hbar}{2m} \left(\psi^* \frac{\partial \psi}{\partial x} - \psi \frac{\partial \psi^*}{\partial x} \right) \Bigg|_{x_1}^{x_2} \tag{1.17}$$

Assuming the particle cannot just disappear, then the change in probability inside $\mathscr{V}$ must be exactly compensated by the flow of probability across the boundary of $\mathscr{V}$. Hence we may define a *probability current density* $j(x, t)$ to be the *negative* of the right-hand side of equation (1.17), so

rate of increase of probability that particle is in $\mathscr{V}$

$$= j(x_1, t) - j(x_2, t)$$

$$= \text{flow into } \mathscr{V} \text{ from left} + \text{flow into } \mathscr{V} \text{ from right}$$

$$= \text{total flow into } \mathscr{V}$$

In three dimensions one defines, analogously,

$$j(r, t) = -\frac{i\hbar}{2m} (\psi^* \nabla \psi - \psi \nabla \psi^*) \tag{1.18}$$

Using the probability density $P(r, t)$ we can construct the average or expectation value of the particle's position by weighting each position r with its associated probability density and integrating

$$\langle r \rangle = \int rP(r, t)\, d\tau = \int \psi^*(r, t)r\psi(r, t)\, d\tau \qquad (1.19)$$

Similarly $\langle f(r) \rangle$ is given by equation (1.19) with r replaced by $f(r)$. The latter form of writing is in anticipation of the more complete treatment of quantum mechanics given in Chapter 6, in which the expectation values of other quantities will be discussed. The use of $\langle\ \rangle$ to indicate an expectation value is common; it forms a part of a system known as Dirac notation which we find useful at a later stage. The next time you encounter it will be in section 6.1.

2

Wave mechanics 1

The study of the behaviour of microscopic systems using Schrödinger's wave equation is a branch of quantum mechanics known as wave mechanics. This approach is inadequate to deal with all problems of interest, but a remarkably wide range of processes can successfully be investigated solely with the theory given at the end of the last chapter. We shall pursue wave mechanics as far as we can in this chapter and the next three, and leave the further, more abstract, developments until Chapter 6.

2.1 TIME-INDEPENDENT SCHRÖDINGER EQUATION

The physics of a particle which moves in a *static* potential $V(r)$ is of particular interest. Classically, the energy of such a particle is conserved. In the Schrödinger equation (1.12) we may, in this special case, separate the r and t variables by substituting $\psi(r,t) = u(r)f(t)$ and then divide by ψ to obtain

$$\frac{1}{u}\left[-\frac{\hbar^2}{2m}\nabla^2 u + V(r)u \right] = \frac{i\hbar}{f}\frac{\partial f}{\partial t} \tag{2.1}$$

In this equation the left-hand side is a function of r only, while the right-hand side is a function of t only. Two functions of different variables can only be equal for all values of their arguments if each is equal to the same constant. Call the constant E. Then

$$-\frac{\hbar^2}{2m}\nabla^2 u(r) + V(r)u(r) = Eu(r) \tag{2.2}$$

and

$$i\hbar \frac{\partial f(t)}{\partial t} = E f(t) \tag{2.3}$$

Equation (2.3) may be integrated immediately. The partial derivative can be replaced by a total derivative because f is a function of t only. Solving, we get $f(t) = \exp(-iEt/\hbar)$. Thus

$$\psi(\mathbf{r}, t) = u(\mathbf{r}) e^{-iEt/\hbar} \tag{2.4}$$

Evidently the time-dependent part of ψ is a pure *phase factor*. Recall from section 1.6 that the position probability density is independent of the phase

$$P = |\psi(\mathbf{r}, t)|^2 \equiv |u(\mathbf{r})|^2 \tag{2.5}$$

Thus although the wave function ψ is time dependent, the observable quantity P is constant in time. Quantum states for which this situation holds are known as *stationary states*.

To determine the meaning of the constant E we can consider the special case of a free particle by putting $V = 0$ in equation (2.2), i.e. $\nabla^2 u = -(2mE/\hbar^2)u$, which has solutions of the form

$$u \propto e^{i\mathbf{k}\cdot\mathbf{r}}, k^2 \equiv |\mathbf{k}|^2 = 2mE/\hbar^2 \tag{2.6}$$

However, de Broglie's relation, equation (1.7), which can be rewritten $p = \hbar k$, relates the wave number k to the magnitude of the momentum p. Using this relation in equation (2.6) we find $E = p^2/m$. Evidently E represents the *energy* of the particle. (When $V \neq 0$, E will be the sum of the kinetic and potential energy.) So in the quantum case, too, energy is conserved. Equation (2.2) therefore reads rather like an energy equation, with the right-hand side representing the total energy E and the left-hand side composed of a term $V(\mathbf{r})u(\mathbf{r})$, representing potential energy, and $(-\hbar^2/2m)\nabla^2 u(\mathbf{r})$, in some way representing kinetic energy.

Equation (2.2) is known as the time-independent Schrödinger equation, and the solutions $u(\mathbf{r})$ are also called wave functions. The full wave function $\psi(\mathbf{r}, t)$ of these stationary states always has the form as shown in equation (2.4), with E being the total energy of the particle. We shall often have occasion to use the one-dimensional form of equation (2.2). This is

$$-\frac{\hbar^2}{2m} \frac{d^2 u(x)}{dx^2} + V(x)u(x) = Eu(x) \tag{2.7}$$

An important type of problem is that in which some potential function V is specified and equation (2.2) is solved for $u(r)$. It may then happen that acceptable solutions for $u(r)$ can only be found for certain restricted values of E.

To be physically acceptable, a wave function must satisfy the following important conditions.

1. ψ is a single-valued function of position and time. If this were not so we should not know which value corresponds to physical reality.

2. ψ is normalizable. This usually means that when $|\psi|^2$ is integrated over all space the answer must be finite. If this were not so we should not be able to use the probability interpretation of $|\psi|^2$ which requires that all probabilities be $\leqslant 1$. In the special case of free particles the normalization condition is sometimes constructed slightly differently (section 4.3).

In addition to our imposing these conditions, solutions of the Schrödinger equation will have the following property.

3. ψ and $\nabla\psi$ will be continuous everywhere except where V has an infinite discontinuity. If this were not so $\nabla^2\psi$ would be infinite at the discontinuity and could not satisfy the Schrödinger equation for finite V. In the idealized case of an infinite jump in V (section 2.2) only ψ need be continuous.

We shall see in the following sections that it is just these restrictions on ψ which lead to the existence of discrete energy levels.

2.2 THE INFINITE SQUARE-WELL POTENTIAL

The simplest illustration of the above ideas is the case of a particle confined to a rigid one-dimensional box of size $2a$. Impenetrable box walls can be modelled by the potential

$$V(x) = 0 \qquad |x| < a$$
$$= \infty \qquad |x| > a$$

As this potential is static we use the time-independent form of the Schrödinger equation, [2.7], which in this case has the form

$$-\frac{\hbar^2}{2m}\frac{d^2u}{dx^2} = Eu \qquad |x| < a$$

with solutions

$$u(x) = A \sin \alpha x + B \cos \alpha x \qquad (2.8)$$

where A and B are constants and $\alpha = (2mE/\hbar^2)^{1/2}$. As the box is impenetrable we must have $u = 0$ for $|x| > a$.

Recalling that u must be continuous, we find at $x = \pm a$

$$\left. \begin{array}{c} A \sin \alpha a + B \cos \alpha a = 0 \\ - A \sin \alpha a + B \cos \alpha a = 0 \end{array} \right\} \qquad (2.9)$$

There are two non-trivial sets of solutions of equations (2.9). Either $A = \cos \alpha a = 0$, or $B = \sin \alpha a = 0$. In either case

$$\alpha = n\pi/2a \qquad (2.10)$$

with $n = 1, 3, 5 \ldots$ or $2, 4, 6 \ldots$ respectively.

This is a crucial step because it is where the energy levels appear. By imposing the boundary conditions (equations (2.9)) on the solution (equation (2.8)) we immediately restrict the parameter α to the discrete set of values (equation (2.10)). Remembering the definition of α, this yields

$$E \equiv E_n = n^2 \pi^2 \hbar^2 / 8ma^2, \quad n = 1, 2, 3 \ldots \qquad (2.11)$$

The energy E, which is a continuous parameter in classical mechanics, is here 'quantized' to certain discrete energy levels. In the classical limit, $\hbar \to 0$, the levels collapse together and discreteness disappears, but for $Ema^2 \approx \hbar^2$ the discreteness will be physically very important. To take an example, if m is the mass of the electron and $a \approx 10^{-10}$ m (about the size of an atom) we find the lowest energy levels spaced out by several electron volts. Clearly we are on the track of the atomic energy levels, which have binding energies of this order. Like atoms, the rigid box has an infinite number of energy levels, but unlike atoms the particle is always bound in the box. Atomic energy levels get closer and closer together like $1/n^2$ as the energy is raised (equation (1.5)), whereas the box levels grow farther apart like n^2. This difference can be traced to the rigidity of the box potential walls (atomic binding weakens with distance).

Another important feature of equation (2.11) concerns the *lowest* energy level $n = 1$ (note that $n = 0$ corresponds to $u = 0$, so occurs with zero probability). This is known as the *ground state* and has energy $\pi^2 \hbar^2 / 8ma^2$, in spite of the fact that $V = 0$ inside the box. Classically there is no reason why both the potential and kinetic energy of the particle cannot be zero. The lowest energy state would therefore have zero total energy E. In quantum mechanics this is impossible for the following reason. If the particle is known to be confined in the region $-a < x < a$ then, according to Heisenberg's uncertainty principle (equation (1.11)), the momentum cannot be fixed to within $\Delta p \approx \hbar/a$, corresponding to a kinetic energy of $(\Delta p)^2/2m \approx \hbar^2/2ma^2$, which is, to within a numerical factor of order unity, the gound state energy found above. Notice that as the box is made large ($a \to \infty$) the ground state energy falls to zero.

More useful information is provided by the wave function (equation (2.8)). First the constants A and B must be determined. This is achieved by applying the normalization condition, equation (1.14). For example, for even n we require

$$\int_{-\infty}^{\infty} |\psi|^2 \, \mathrm{d}x = \int_{-a}^{a} |u|^2 \, \mathrm{d}x = A^2 \int_{-a}^{a} \sin^2(n\pi x/2a) \, \mathrm{d}x = 1$$

from which we find $A^2 = 1/a$. Similarly $B^2 = 1/a$. Thus

$$\left. \begin{array}{ll} u(x) = a^{-1/2}\sin(n\pi x/2a) & n \text{ even} \\ u(x) = a^{-1/2}\cos(n\pi x/2a) & n \text{ odd} \end{array} \right\} \qquad (2.12)$$

The probability density $|u|^2$ is shown for $n = 1$ and 2 in Fig. 2.1. The wave function is reminiscent of the classical problem of standing waves on a stretched string with fixed endpoints, with the stationary states corresponding to the normal modes of vibration. Just as there

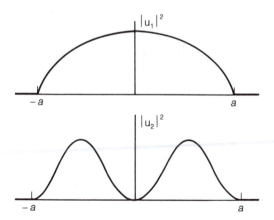

Fig. 2.1 Position probability densities for the ground state ($n = 1$) and first excited state ($n = 2$) of a particle trapped in a one-dimensional impenetrable box of length $2a$. Note the symmetry about the origin.

are discrete frequencies of vibration on the string, so there are discrete energy levels here (the wave frequency ω being $E/\hbar$). Notice that the probability of finding the particle near some point x is far from uniform inside the box. In all cases the particle tends to shy away from the walls of the box, while in the even n cases it avoids the centre. However, as n becomes very large, $|u|^2$ wiggles rapidly and, on a coarse scale, looks uniform. This is the classical limit where the particle is equally likely to be found near any point in the box. By symmetry the expectation value of x vanishes for all n. For example, for even n

$$\langle x \rangle = \frac{1}{a} \int_{-a}^{a} x \sin^2(n\pi x/2a)\,\mathrm{d}x = 0 \tag{2.13}$$

2.3 DEGENERACY AND PARITY

An important property of the energy levels considered in the previous section is that there is a one-to-one correspondence between the stationary-state energies E_n and the states (as labelled by the index n or the wave functions u_n). This may not always be the case. Consider, for example, the three-dimensional box

$$V = 0 \qquad |x| < a_1, \quad |y| < a_2, \quad |z| < a_3$$
$$= \infty \qquad \text{otherwise}$$

Repeating the treatment of the previous example, we must solve equation (2.2) in Cartesian coordinates. Inside the box

$$-\frac{\hbar^2}{2m}\left(\frac{\partial^2}{\partial x^2} + \frac{\partial^2}{\partial y^2} + \frac{\partial^2}{\partial z^2}\right)u(x, y, z) = Eu(x, y, z)$$

which has the solutions of the form

$$u(x, y, z) = (A_1 \sin \alpha_1 x + B_1 \cos \alpha_1 x)(A_2 \sin \alpha_2 y$$
$$+ B_2 \cos \alpha_2 y)(A_3 \sin \alpha_3 z + B_3 \cos \alpha_3 z)$$

with

$$E = (\alpha_1^2 + \alpha_2^2 + \alpha_3^2)\hbar^2/2m$$

Imposing the boundary conditions that $u = 0$ at $|x| = a_1, |y| = a_2, |z| = a_3$ forces

$$\alpha_1 = n_1\pi/2a_1, \quad \alpha_2 = n_2\pi/2a_2, \quad \alpha_3 = n_3\pi/2a_3$$

where n_1, n_2 and n_3 are positive integers. The energy levels are therefore

$$E_{n_1 n_2 n_3} = \left(\frac{n_1^2}{a_1^2} + \frac{n_2^2}{a_2^2} + \frac{n_3^2}{a_3^2} \right) \frac{\hbar^2 \pi^2}{8m}$$

In the special case of the cube, $a_1 = a_2 = a_3 = a$, say,

$$E_{n_1 n_2 n_3} = (n_1^2 + n_2^2 + n_3^2)\hbar^2 \pi^2 / 8ma^2 \tag{2.14}$$

In this case different sets of numbers (n_1, n_2, n_3) can give rise to the same energy E. For example $(1, 2, 3)$, $(2, 1, 3)$, $(3, 1, 2)$, $(1, 3, 2)$, $(2, 3, 1)$ and $(3, 2, 1)$ all have the same E. There are thus six different quantum stationary states all with energy $14\pi^2\hbar^2/8ma^2$. This six-fold degeneracy would not have happened if $a_1 \neq a_2 \neq a_3$. Evidently the occurrence of *degeneracy* is closely connected with the concept of *symmetry*.

Returning to the one-dimensional case, the wave functions (equation (2.12)) have the important property that they are either even or odd under reflection $(x \to -x)$

$$u(-x) = u(x) \qquad \text{for } n = 1, 3, 5 \ldots$$
$$u(-x) = -u(x) \quad \text{for } n = 2, 4, 6 \ldots$$

This even-ness or odd-ness is referred to as the *parity* of the state.

A non-degenerate stationary-state wave function will have a definite even or odd parity whenever the potential $V(x)$ is symmetric under reflection

$$V(-x) = V(x) \tag{2.15}$$

This is easily proved. Change x to $-x$ in equation (2.7):

$$-\frac{\hbar^2}{2m} \frac{\mathrm{d}^2 u(-x)}{\mathrm{d}x^2} + V(x)u(-x) = Eu(-x) \tag{2.16}$$

where we have replaced $V(-x)$ by $V(x)$ in accordance with equation (2.15). Equation (2.16) is identical to the Schrödinger equation for $u(x)$. As the system is non-degenerate there can only exist one stationary state for each value of E, so $u(x)$ and $u(-x)$ must represent the same solution. Thus $u(x)$ and $u(-x)$ can differ by at most a constant factor

$$u(x) = \varepsilon u(-x) \tag{2.17}$$

However, reversing the sign of x,

$$u(-x) = \varepsilon u(x). \tag{2.18}$$

From equations (2.17) and (2.18) we conclude

$$u(x) = \varepsilon^2 u(x)$$

implying $\varepsilon = \pm 1$ and, as claimed, $u(-x) = \pm u(x)$.

Symmetric potentials have the important property that the expectation value of the position of the particle in the potential always vanishes:

$$\langle x \rangle = \int_{-\infty}^{\infty} |u(x)|^2 x \, dx = 0 \qquad (2.19)$$

The reason is simple. As $u(x)$ must be either an even or odd function of x, $|u|^2$ is even. Hence $x|u|^2$ is odd, and the contribution to the integral for $x > 0$ exactly cancels that from $x < 0$. An example of equation (2.19) is given by equation (2.13).

2.4 FINITE SQUARE WELL

An important new feature of quantum systems emerges when we consider the more realistic case of a finite square-well potential

$$
\begin{aligned}
V(x) &= V_0 > 0 \qquad |x| > a \\
&= 0 \qquad\quad |x| < a
\end{aligned}
\qquad (2.20)
$$

Once again the stationary states are found by solving equation (2.7), which in this case becomes

$$-\frac{\hbar^2}{2m}\frac{d^2u}{dx^2} = Eu \qquad\qquad |x| < a$$

$$-\frac{\hbar^2}{2m}\frac{d^2u}{dx^2} + V_0 u = Eu \qquad |x| > a$$

Introducing constants $\alpha = (2mE/\hbar^2)^{1/2}$ and $\beta = [2m(V_0 - E)/\hbar^2]^{1/2}$, and assuming $E < V_0$, these equations reduce to the simple form

$$\frac{d^2u}{dx^2} = -\alpha^2 u \qquad |x| < a$$

$$\frac{d^2u}{dx^2} = \beta^2 u \qquad |x| > a$$

The functional form of the solutions for $|x| < a$ is the same as for the infinite square well, equation (2.8). For $|x| > a$

$$u = Ce^{-\beta x} + De^{\beta x} \qquad (2.21)$$

where C and D are constants.

To save labour one can make use of the fact that, as the potential is symmetric in x, solutions will have definite even or odd parity. We then need only solve in the half-space $x > 0$ and deduce the solution for $x < 0$ by reflection. For $x > 0$, $e^{\beta x}$ is unacceptable because $\int |u|^2 \, dx$ will blow up as $x \to \infty$ implying that u cannot be normalized. Therefore $D = 0$.

For even parity solutions we must have $A = 0$ in equation (2.8). Thus

$$u = B \cos \alpha x \qquad 0 < x < a$$
$$= Ce^{-\beta x} \qquad x > a$$

Continuity in u and du/dx at $x = a$ then requires

$$B \cos \alpha a = Ce^{-\beta a} \qquad (2.22)$$

$$- B\alpha \sin \alpha a = - C\beta e^{-\beta a} \qquad (2.23)$$

respectively. Dividing (2.23) by (2.22) gives

$$\tan \alpha a = \beta/\alpha \qquad (2.24)$$

Similarly, for odd parity solutions $B = 0$ and one finds

$$\cot \alpha a = - \beta/\alpha \qquad (2.25)$$

The positioning of the energy levels is determined by solutions of equations (2.24) and (2.25). No solutions exist in terms of elementary functions, and numerical or graphical techniques are necessary. The qualitative features are easily deduced. Consider, for example, equation (2.25). If β/α is plotted as a function of E, then the energy levels will be given by the values at which this curve intersects the graph of $-\cot \alpha a$. It is convenient to introduce the constant $\gamma = (2mV_0 a^2/\hbar^2)^{1/2}$ and the variable $x = \gamma(E/V_0)^{1/2}$. Then

$$\beta/\alpha = [(\gamma/x)^2 - 1]^{1/2}$$

and

$$- \cot \alpha a = - \cot x$$

Figure 2.2 shows graphs of $-\cot x$ and $[(\gamma/x)^2 - 1]^{1/2}$ for three values of γ. Note that $x \geqslant 0$, and for real values of x we must have $[(\gamma/x)^2 - 1]^{1/2} \geqslant 0$ too. The curves $[(\gamma/x)^2 - 1]^{1/2}$ reach the x axis

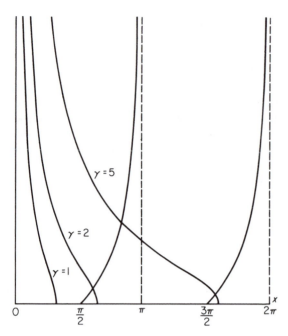

Fig. 2.2 Graphs of $-\cot x$ and $[(\gamma/x)^2 - 1]^{1/2}$ are shown. As γ is increased so the latter curves reach out farther to the right. The points of intersection determine the values of E which correspond to the allowed energy levels.

at $x = \gamma$. Clearly if $\gamma < \pi/2$ these curves do not intersect $-\cot x$ at all. There are no energy levels. For γ between $\pi/2$ and $3\pi/2$ there will be one energy level, given by the value of E at which the curve cuts the first branch of $-\cot x$. For γ between $3\pi/2$ and $5\pi/2$ the curve $[(\gamma/x)^2 - 1]^{1/2}$ intersects two branches of $-\cot x$; there are thus two allowed energy levels. As the value of γ is increased so the curve $[(\gamma/x)^2 - 1]^{1/2}$ intersects more and more branches of $-\cot x$. There are, of course, an infinite number of these branches. The even parity solutions (equation (2.24)) may be treated similarly.

The number of energy levels is seen, therefore, to be determined entirely by the parameter γ. This contains the factor $(V_0 a^2)^{1/2}$, which can be taken as a measure of the 'strength' of the well. If V_0 is increased, the well gets deeper; if a is increased, it gets wider. In both cases γ increases and more energy levels appear. The reason why there are no levels for $\gamma < \pi/2$, i.e. for $V_0 a^2 < \pi^2 \hbar^2/8m$, can be traced to the Heisenberg uncertainty principle. Being confined to a region of size $\approx a$ the particle experiences a momentum uncertainty

$\Delta p \approx \hbar/a$, which for an otherwise free particle corresponds to an energy $(\Delta p)^2/2m \approx \hbar^2/2ma^2$. If this is greater than the depth of the well V_0 then no bound state will be possible. The well will not be strong enough to overcome the irreducible quantum agitation of the particle. Such a circumstance occurs if $\hbar^2/2ma^2 \approx V_0$ or if $V_0 a^2 \approx \hbar^2/2m$, which is a crude approximation to the condition found above.

In Fig. 2.3 the wave functions for the three lowest energy levels are shown for the case $V_0 a^2 = \pi^2 \hbar^2/m$. An important feature of all these functions is the exponential decline of $u(x)$ in the region $|x| > a$ (equation (2.21)). In the limit of the infinite well V_0 (hence β) $\rightarrow \infty$ and these exponential 'wings' to the wave function pinch off. The infinite barrier is impenetrable. The finite well, however, permits the particle to penetrate the wall for a short distance. There is a finite probability that the particle will be found *outside* the limits $|x| = a$ of the well, even though classically the energy $E < V_0$ is insufficient for the particle to reach this region.

Finally we examine the case $E > V_0$. Classically such a particle is not bound by the well. We find β is imaginary, so we put $\beta = i\kappa$. Equation (2.21) becomes

$$u(x) = C \sin \kappa x + D \cos \kappa x \qquad |x| > a$$

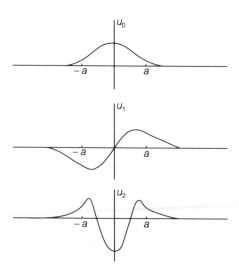

Fig. 2.3 Wave functions for the first three energy levels of the finite square-well system. The position probability densities are obtained by squaring these functions. Note that the functions are non-zero, but rapidly declining, in the classically inaccessible region $|x| > a$.

where C and D are new constants. Continuity in u and du/dx at $x = \pm a$ and the normalization condition then fixes all the four constants A, B, C and D, but there is now no restriction on α or κ or, hence, on E. In the region $E > V_0$, then, the energy levels form a continuum and the wave functions do not decline at large $|x|$; the particle is thus unbound and has an appreciable probability of being found far from the well.

2.5 BOUNCING BALL

Consider the problem of a ball which bounces vertically on a rigid horizontal plane. The potential energy may be modelled by

$$\left.\begin{array}{ll} V = mgx & x > 0 \\ \;\;= \infty & x < 0 \end{array}\right\} \tag{2.26}$$

where x is the vertical height above the plane and g is the acceleration due to gravity (Fig. 2.4). Equation (2.7) becomes

$$-\frac{\hbar^2}{2m}\frac{d^2u}{dx^2} + mgxu = Eu \tag{2.27}$$

Putting $x = az + b$, where $a^3 = \hbar^2/2m^2g$ and $b = E/mg$, equation (2.27) reduces to

$$\frac{d^2u}{dz^2} - zu = 0 \tag{2.28}$$

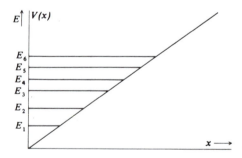

Fig. 2.4 The potential $V(x)$ for a ball bouncing on a rigid horizontal plane is shown, with x representing the vertical height above the plane. The energy levels are superimposed on the diagram. Notice that the spacing slowly diminishes. In the classical limit (n very large) the discrete levels are so densely packed that they approximate to a continuum.

which is a form of Bessel's equation. The solutions are Bessel functions of order 1/3, more commonly known as Airy functions, and denoted by Ai(z) and Bi(z). Of these two functions only the Ai(z) are acceptable as wave functions because the Bi(z) cannot be normalized. Thus $u(x) \propto \mathrm{Ai}(z) = \mathrm{Ai}[(x-b)/a]$. The boundary condition $u = 0$ at $x = 0$ then requires $\mathrm{Ai}(-b/a) = 0$, and the energy levels are given by the roots of this equation. The function $\mathrm{Ai}(-b/a)$ is an oscillatory function of b/a and is illustrated in Fig. 2.5; the roots can be found numerically and the first ten occur at the values of b/a given to three decimal places in the left-hand column of Table 2.1. For comparison, the right-hand column gives results calculated from an analytic asymptotic form $[3\pi(n-1/4)/2]^{2/3}$ which is strictly accurate only as $n \to \infty$.

Thus we see that the asymptotic approximation is accurate to better than 1% even for the first root, and improves rapidly for the higher roots. We conclude that the following expression, though not quite exact, is always accurate to better than 1%:

$$\frac{E_n}{(m\hbar^2 g^2/2)^{1/3}} = [3\pi(n-1/4)/2]^{2/3} \tag{2.29}$$

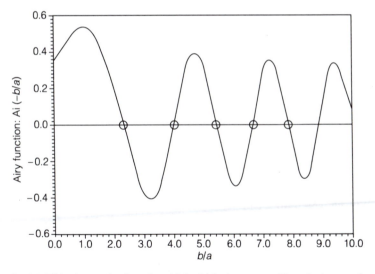

Fig. 2.5 This shows the function Ai ($-b/a$), the acceptable solution to the Schrödinger equation (2.27), where $a^3 = \hbar^2/2m^2 g$ and $b = E/mg$. The ringed points are some of the roots of the equation Ai($-b/a$) = 0. Numerical values of the first ten roots are given in Table 2.1.

Table 2.1 *The first ten roots of the function* $Ai(-b/a)$

n	b/a (true)	b/a (approximate)
1	2.338	2.320
2	4.088	4.082
3	5.521	5.517
4	6.787	6.784
5	7.944	7.942
6	9.023	9.021
7	10.040	10.039
8	11.009	11.008
9	11.936	11.935
10	12.829	12.828

Fig. 2.6 This shows plots of the first three wave functions (for $n = 1, 2$ and 3) $u(x) = Ai[(x - b)/a]$ versus x/a.

where $n = 1, 2, 3 \ldots$. For large n the levels go like $n^{2/3}$. This may be compared with n^2 for the infinite square well. The closer spacing of levels is connected with the shallower rate of climb of the potential (equation (2.26)).

Figure 2.6 shows the forms of the unnormalized wave functions $u(x)$ for the first three energy levels. It is of interest to take an example and to calculate the magnitudes. For example if we take m to be equal to the mass of a helium atom, then the distance a is $a = 2.33\,\mu m$ and the energy of the state with $n = 1$ is $E_1 = 3.57 \times 10^{-31}\,J$ (very small by comparison with room-temperature thermal energies of about $4.14 \times 10^{-21}\,J$).

3

One-dimensional delta wells

In this chapter, we apply some of the ideas from the last chapter to a special case which can fairly readily be extended to describe an important area, namely, the behaviour of electrons in a crystal lattice.

3.1 THE SINGLE ONE-DIMENSIONAL DELTA WELL

The starting point is section 2.4 which dealt with a particle in a single one-dimensional square well of finite depth. We consider a particle with mass m and total energy E moving under the influence of the one-dimensional finite square well illustrated in Fig. 3.1.

For reasons which will become apparent, we focus attention on the ground-state solution of equation (2.7) with the potential shown in Fig. 3.1. In the notation section 2.4, that solution for the region inside the well is

$$u = B \cos \alpha x \qquad (3.1)$$

where

$$\alpha = \left(\frac{2mE}{\hbar^2} \right)^{1/2} \qquad (3.2)$$

For the region on the left-hand side (LHS) outside the well the solution is

$$u_{\text{LHS}} = C \exp(+ \beta x) \qquad (3.3a)$$

and that for the region on the right-hand side (RHS) outside the well is

$$u_{\text{RHS}} = C \exp(- \beta x) \qquad (3.3b)$$

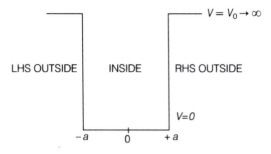

Fig. 3.1 This shows the potential $V(x)$ as a function of position x on a line. Inside the well, the potential is zero; outside it is V_0.

where

$$\beta = \left[\frac{2m(V_0 - E)}{\hbar^2} \right]^{1/2} \tag{3.4}$$

The number of bound states depends on the magnitude of the dimensionless parameter

$$\gamma \equiv (2mV_0 a^2/\hbar^2)^{1/2} \tag{3.5}$$

$$= a(\alpha^2 + \beta^2)^{1/2} \tag{3.6}$$

introduced in section 2.4. In fact there is a simple way of working out from a consideration of Fig. 2.2 just how many bound states there are. If the number of bound states is denoted by N, then N is that integer which satisfies

$$(N-1)\pi < 2\gamma \leqslant N\pi \tag{3.7}$$

Now we modify the square well to the limiting form of a delta well by allowing $a \rightarrow 0$ and $V_0 \rightarrow \infty$ in such a way that the product $\delta \equiv V_0 a$ remains finite. Equation (3.5) shows that in this limit $\gamma \rightarrow 0$ and according to equation (3.7) $N = 1$ which means that only one bound state exists. The appearance of the wave function for this state is given by equations (3.3a), (3.1) and (3.3b) with $C = B$, and is illustrated in Fig. 3.2; it has even parity.

The magnitudes of αa and βa are easily deduced. First, from the defining equation (3.2), we see that since $E < V_0$, αa must be less than $(2m\delta a/\hbar^2)^{1/2}$ and so must be approaching zero because $a \rightarrow 0$, all other terms being finite. We can next use equation (2.24),

$$\tan \alpha a = \beta/\alpha \tag{3.8}$$

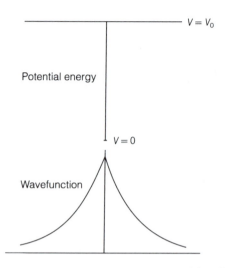

$V = V_0$

Potential energy

$V = 0$

Wavefunction

Fig. 3.2 The upper part of this figure shows a potential well like that in Fig. 3.1, except that it has zero width. The lower part shows the ground-state wave function as a function of distance from the well.

in its limiting form

$$(\alpha a)^2 = \beta a \tag{3.9}$$

which is appropriate because we can approximate $\tan \alpha a$ by αa. Combining equations (3.6) and (3.9) now gives

$$(\alpha a)^4 + (\alpha a)^2 - \gamma^2 = 0 \tag{3.10}$$

Remembering that αa is vanishingly small we neglect $(\alpha a)^4$ in comparison with $(\alpha a)^2$ and conclude that

$$(\alpha a)^2 = \gamma^2 \tag{3.11}$$

so that

$$a\alpha^2 = 2mV_0 a/\hbar^2 = 2m\delta/\hbar^2 \tag{3.12}$$

Finally from equation (3.9) and the defining equation (3.4) for β we deduce that

$$\beta^2 = 2m(V_0 - E)/\hbar^2 = (a\alpha^2)^2 = (2m\delta/\hbar^2)^2 \tag{3.13}$$

so that

$$(V_0 - E) = 2m\delta^2/\hbar^2 \tag{3.14}$$

It is interesting to note that the right-hand side of this equation is positive and finite even though V_0 approaches infinity. This tells us that E also approaches infinity while remaining a finite amount below V_0.

The problem of the delta well is thus solved. There is, however, an alternative method which turns out to be much more flexible when arrays of two or more wells are considered. The starting point of this alternative method is the boundary condition that the gradient of the wave function must be continuous at both edges of the well, and this implies that the gradient change across the well must be the same whether one uses the internal or external wave functions. For the internal region, the time-independent Schrödinger equation (TISE) is

$$\frac{d^2 u}{dx^2} = \alpha^2 u \tag{3.15}$$

Now the infinitesimal width of the well means that inside we may take u to be essentially a constant (and equal to the value immediately outside on either side). Integrating equation (3.15) is therefore easy:

$$\left(\frac{du}{dx}\right)_{RHS} - \left(\frac{du}{dx}\right)_{LHS} = -\alpha^2 u_w \int_{-a}^{+a} dx \tag{3.16}$$

$$= -\frac{4m\delta}{\hbar^2} u_w \tag{3.17}$$

where u_w is the constant value of u at the well and we have used equation (3.12). This gradient change can obviously be equated to the gradient difference which follows easily from the outside solutions given in equation (3.13), i.e. $-2\beta^{3/2}$. The result is the same as equation (3.10) and so has the same answer, namely $(V_0 - E) = 2m\delta^2/\hbar^2$. The advantage of this alternative method is that it is more readily extended to the triple one-dimensional delta well which we now consider, and further to multiple one-dimensional delta wells.

3.2 THE ONE-DIMENSIONAL TRIPLE DELTA WELL

After considering the single delta well, it is instructive to turn to the double or triple delta well. We leave the double well as an exercise for the reader, but consider the bound states in the triple well in some detail here. Further, in Appendix A, we shall also say a little about the infinite array of delta wells because of its relevance to the motion of electrons in crystal lattices, but will omit the natural further extension into three dimensions.

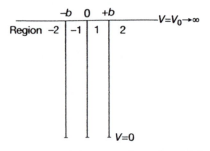

Fig. 3.3 This shows a group of three one-dimensional delta wells, each like the one illustrated in Fig. 3.1 The central well is situated at $x = 0$, while the side wells are situated at $x = \pm b$.

Consider the array of three one-dimensional delta wells illustrated in Fig. 3.3. This potential configuration can have up to three allowed states, depending on the well separation r. Since the potential is symmetrical about $r = 0$, the wave functions must have even or odd parity. Two of them have even parity while the third has odd parity. The wavefunctions in the regions 1 and 2 are readily found as solutions of the TISE

$$\frac{d^2 u}{dx^2} = \beta^2 u \tag{3.18}$$

We take, respectively,

$$u_1 = A \exp(\beta x) + B \exp(-\beta x) \tag{3.19}$$

and

$$u_2 = C \exp(-\beta x) \tag{3.20}$$

(Because of the symmetry of the potential, there is no need to treat regions -1 and -2 separately.) Following the logic of the arguments above, the boundary conditions are that at $x = b$,

$$u_1 = u_2 \tag{3.21}$$

and

$$\frac{du_2}{dx} - \frac{du_1}{dx} = -\frac{4m\delta}{\hbar^2} u_2 \tag{3.22}$$

while at $x = 0$,

either

$$u_1 = 0 \quad \text{(odd parity)} \tag{3.23a}$$

or

$$2\frac{du_1}{dx} = -\frac{4m\delta}{\hbar^2} u_1 \quad \text{(even parity)} \tag{3.23b}$$

Application of these three boundary conditions, together with the normalization condition, now leads to a complete determination of for four constants A, B, C, and the energy $(E - V_0)$ all in terms of the separation b. It is convenient, though not essential, to work with dimensionless units:

$$s \equiv a\alpha^2 b = \frac{2m\delta}{\hbar^2} b \qquad (3.24)$$

and

$$\varepsilon \equiv \frac{\hbar^2}{2m\delta^2}(E - V_0) = -\left(\frac{\beta}{a\alpha^2}\right)^2 \qquad (3.25)$$

It is well to bear in mind that ε is negative (and therefore $-\varepsilon$ is positive) in the equations below. This makes physical sense since we are dealing with bound states with energies less than V_0. In Fig. 3.4 the normalized wave functions are plotted for the case of $s = 2$. The corresponding allowed energies are indicated in the figure, and they have been derived numerically from the equations which follow from

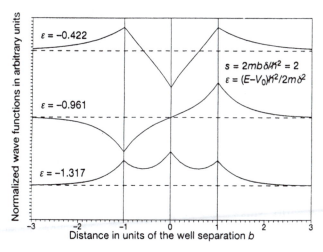

Fig. 3.4 The three bound-state wave functions for the triple well shown in Fig. 3.3. The wave functions are normalized, and distances are in the dimensionless units x/b. The value of the dimensionless separation parameter s is taken as 2, and the corresponding values of the dimensionless energy parameter ε are shown against the curves. The effect of changing the separation is shown in Fig. 3.5.

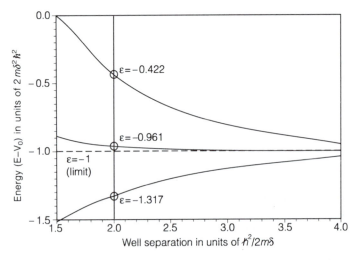

Fig. 3.5 When the well separation is varied, the energies of the three bound states change. This shows how the dimensionless energy parameter ε varies with the dimensionless separation parameter s. The circled points are for $s = 2$, the case for which Fig. 3.4 applies.

the boundary conditions stated above. The reader may care to verify these equations:

$$s = -\frac{1}{2(-\varepsilon)^{1/2}}\ln[1-(-\varepsilon)^{1/2}] \quad \text{(odd parity)} \qquad (3.26a)$$

with allowed values of ε in the range -1 to 0, or

$$s = -\frac{1}{2(-\varepsilon)^{1/2}}\ln\left\{\frac{[1-(-\varepsilon)^{1/2}]^2}{1+(-\varepsilon)^{1/2}}\right\} \quad \text{(even parity)} \qquad (3.26b)$$

with one set of allowed values of ε in the range -1 to 0, and a second with allowed values of ε in the range -9 to -1. We note that the limiting case of $s = 0$ (overlap of the three wells) corresponds to even parity with $\varepsilon = -9$, while $s \to \infty$ (infinite separation) is the limiting case in which all three cases give $\varepsilon = -1$. Figure 3.5 shows the dependence of ε on s for the three states illustrated (for $s = 2$) in Fig. 3.4.

Further development of these ideas can be found in Appendix A, where large numbers of wells are considered. This is directly relevant to the behaviour of conduction electrons in crystal lattices.

4

Wave mechanics 2

In this chapter energy levels for more realistic systems will be obtained. We shall then look at a different class of problems involving the scattering of free particles from a potential. Finally some remarks will be made about many-particle systems.

4.1 SIMPLE HARMONIC OSCILLATOR

This example is important because many physical systems undergoing small disturbances behave like simple harmonic oscillators, e.g. crystal lattices, diatomic molecules. We restrict the treatment to one space dimension.

Classically the system is described by the static potential

$$V = \tfrac{1}{2}Kx^2 \tag{4.1}$$

where K is a constant. This describes a force $-Kx$, for which Newton's second law

$$m\frac{d^2x}{dt^2} = -Kx \tag{4.2}$$

has oscillating solutions $A \sin(\omega t + \varphi)$, with $\omega = (K/m)^{1/2}$, A denoting the amplitude of oscillation, and φ a phase angle determined simply by the position of the mass at $t = 0$. After considering the quantum case, we shall have an expression for the probability of finding the mass m within a range dx at x. But, what is the answer in classical mechanics? Imagine the mass oscillating unobserved in a dark room. At some arbitrary instant you switch on the light. The question is, what is the probability $P(x)dx$ of finding the mass m within a range dx at x? Since $x = A\sin(\omega t + \varphi)$, it follows that

$dx = \omega A \cos(\omega t + \varphi) \, dt = \omega(A^2 - x^2)^{1/2} \, dt$. Obviously the probability of observing the mass within dx is proportional to the time dt that it spends there. So $P(x) \, dx$ is proportional to $dx/\omega(A^2 - x^2)^{1/2}$. The constant of proportionality comes from requiring the integral of $P(x) \, dx$ to be unity over, say, one half-cycle as x goes from $-A$ to $+A$. This leads easily to the classical result

$$P(x) = \frac{1}{\pi(A^2 - x^2)^{1/2}} = \frac{1}{\pi(2E/m\omega^2 - x^2)^{1/2}} \qquad (4.3)$$

where we have used the convenient substitution $E = \frac{1}{2}m\omega^2 A^2$. We shall return to this expression after dealing with the quantum case, and it is shown in Fig. 4.1 for comparison with the quantum case.

In the quantum case we substitute the potential, equation (4.1), into the Schrödinger equation (2.7) to obtain

$$\frac{d^2 u}{dz^2} + (2\varepsilon - z^2)u = 0 \qquad (4.4)$$

where

$$z = (m\omega/\hbar)^{1/2}x, \quad \varepsilon = E/\hbar\omega \qquad (4.5)$$

As usual, one expects solutions of equation (4.4) to show rapid decline as $z \to \pm\infty$. Inspection of the asymptotic form, i.e. $z^2 \gg \varepsilon$, shows $u \sim \exp(-z^2/2)$ is a solution in this region. This therefore suggests general solutions of the form $F(z)\exp(-z^2/2)$, where F is a polynomial. Substituting this form into equation (4.4) yields

$$\frac{d^2 F}{dz^2} - 2z\frac{dF}{dz} + (2\varepsilon - 1)F = 0 \qquad (4.6)$$

(There might also exist non-polynomial, i.e. infinite series, solutions of equation (4.6). This is actually so, but they all have unphysical divergent behaviour at large z.)

Suppose the leading term of F is z^n. This contributes

$$n(n-1)z^{n-2} - 2nz^n + (2\varepsilon - 1)z^n \qquad (4.7)$$

to the left-hand side of equation (4.6). The coefficient of z^n must vanish to comply with equation (4.6) and as lower-order terms in the polynomial F only contribute to z^{n-1}, or lower powers, we demand from equation (4.7) that

$$\varepsilon = n + \tfrac{1}{2} \qquad n = 0, 1, 2, 3, \ldots \qquad (4.8)$$

It follows from equation (4.5) that the energy E is restricted to discrete levels given by

$$E_n = (n + \tfrac{1}{2})\hbar\omega \tag{4.9}$$

These levels have the interesting property that they are equispaced and the classical frequency ω is related to E in the same way as the photon relation (equation (1.1)). This is no coincidence. The electromagnetic field may be treated as a collection of simple harmonic oscillators.

The ground state has energy $\tfrac{1}{2}\hbar\omega$ which, as in previous examples, is above the classical minimum $(E = 0)$. The ground state wave function u_0 is given by $n = 0$ in which case F is constant, so $u_0 \propto \exp(-z^2/2)$. Applying the normalization condition gives a Gaussian function

$$u_0 = (m\omega/\pi\hbar)^{1/4} \exp(-m\omega x^2/2\hbar) \tag{4.10}$$

The expectation values of x and V for the ground state, using equation (1.19) and (4.10), are

$$\langle x \rangle = \int_{-\infty}^{\infty} u_0^2 x \, dx = 0 \tag{4.11}$$

$$\langle V \rangle = \tfrac{1}{2}K \int_{-\infty}^{\infty} u_0^2 x^2 \, dx = \tfrac{1}{2}E_0 \tag{4.12}$$

It follows from equation (4.12) that the expectation value of the kinetic energy $\langle E_0 \rangle - \langle V \rangle$ is $\tfrac{1}{2}E_0$ also. As in the classical case, the average kinetic and potential energies of a simple harmonic oscillator are the same. This remains true for the excited levels.

To construct the excited $(n \geqslant 1)$ wave functions one must substitute a full polynomial for F in equation (4.6) and equate the coefficients of all the powers (not just z^n) to zero. One obtains the following set of functions for F (to within an overall constant) $1, 2z, 4z^2 - 2,$ $8z^3 - 12z, \ldots$ for $n = 0, 1, 2, 3, \ldots$. These functions are well known to mathematicians as Hermite polynomials, denoted $H_n(z)$. (Equation (4.6) is Hermite's equation.) The $H_n(z)$ may all be generated by successive operations as follows:

$$H_n(z) = (-1)^n e^{z^2} \frac{d^n}{dz^n} e^{-z^2} \tag{4.13}$$

The normalization factors may be evaluated using the properties of Hermite polynomials (exercises 4.4–7). One obtains the wave functions

$$u_n(x) = (\alpha/\pi^{1/2} 2^n n!)^{1/2} H_n(\alpha x) e^{-\alpha^2 x^2/2} \tag{4.14}$$

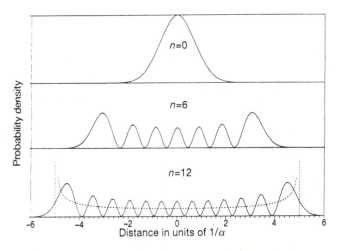

Fig. 4.1 The wave functions $u_0(\alpha x)$, $u_6(\alpha x)$ and $u_{12}(\alpha x)$ are shown. It can be seen that with $n = 12$ the shape has some resemblance to the classical result given in equation (4.4) and shown by a dashed line, although the classical result has no ripples. The wave function is not precisely zero outside the classical amplitude, but falls away in an exponential fashion. The shape of the wave function corresponding to the ground state with $n = 0$ bears no resemblance to the classical result, since it has a maximum, rather than a minimum, at $x = 0$. We conclude that, for large n, the classical result is approached.

with $\alpha = (m\omega/\hbar)^{1/2}$. The functions $u_0(\alpha x)$, $u_6(\alpha x)$, and $u_{12}(\alpha x)$ are illustrated in Fig. 4.1 and, in the latter case, compared with the classical result of equation (4.4) with E_{12} taken as $(2 \times 12 + 1)\hbar\omega = 25\hbar\omega$.

It is instructive to reproduce these results using a different method which, besides being elegant, introduces some useful concepts. Equation (4.4) may be rewritten in either of the following two ways

$$a^- a^+ u = (-2\varepsilon - 1)u \tag{4.15}$$

$$a^+ a^- u = (-2\varepsilon + 1)u \tag{4.16}$$

where

$$a^{\pm} \equiv \frac{\mathrm{d}}{\mathrm{d}z} \mp z \tag{4.17}$$

are known as step-up $(+)$ and step-down $(-)$ operators (for reasons which will soon emerge).

Operating from the left in equation (4.16) with a^-

$$a^- a^+ a^- u = (-2\varepsilon + 1)a^- u$$

or

$$a^- a^+ u' = (-2\varepsilon' - 1)u' \qquad (4.18)$$

where $u' = a^- u$ and $\varepsilon' = \varepsilon - 1$. However, equation (4.18) is just equation (4.15) for u' and ε'. Hence, given the solution u with energy parameter ε, we have discovered a simple procedure for generating another solution u' with lower energy $\varepsilon' = \varepsilon - 1$. Successive applications of this procedure enables us to *step down* level by level. The level spacing in ε is always 1, corresponding to steps of $\hbar\omega$ in E (equation (4.5)). Thus we have already reproduced the result that the energy levels are equispaced by $\hbar\omega$.

The wave functions can all be generated from each other by stepping once we have one to start with. To get the first one note that we cannot step down for ever, for eventually the ground state will be reached . Now a perfectly satisfactory solution of equation (4.18) is $u' = 0$. If we pick this solution as our ground state then we see immediately that the stepping will indeed stop, because $a^- 0 = 0$, and however often we act (step) with a^- thereafter we will still keep getting zero. Hence the ground state wave function must be a solution of $u' = 0$, or

$$u' \equiv a^- u_0 \equiv \left(\frac{\mathrm{d}}{\mathrm{d}z} + z \right) u_0 = 0 \qquad (4.19)$$

which solves immediately to give $u_0 = \text{const.} \times \exp(-z^2/2)$. This is our previous result (equation (4.10)). Note that the ground state wave function u_0 can be defined as that state which is 'annihilated' by a^-, i.e. made to vanish under operation by a^-. Furthermore, substituting equation (4.19) into equation (4.16) implies that $2\varepsilon - 1 = 0$, or $\varepsilon = \frac{1}{2}$. We have therefore recovered the ground state energy $\frac{1}{2}\hbar\omega$ and hence, recalling the equal level steps of $\hbar\omega$, the entire energy spectrum (equation (4.9)).

Having deduced u_0 we can now generate all the other u_n by stepping. As u_0 is at the bottom of the 'ladder' we must do this by stepping up rather than down, which we do by interchanging a^+ and a^-. Thus in place of equation (4.18) we get

$$a^+ a^- u'' = (-2e'' + 1)u'' \qquad (4.20)$$

where $\varepsilon'' = \varepsilon + 1$ and $u'' = a^+ u$ is also a solution (equation (4.16)). So starting with u_0 we can get u_1 from $a^+ u_0, u_2$ from $a^+ u_1$, etc. In this way one essentially reproduces the operation equation (4.13), generating all the Hermite polynomials in sequence.

4.2 HYDROGEN ATOM

The full treatment of this problem will have to await the details of angular momentum to be dealt with in Chapter 7, but we give here a restricted treatment by neglecting the angular variables.

We assume the electron moves radially in a fixed central potential of the Coulomb form

$$V(r) = -\frac{e^2}{4\pi\varepsilon_0 r} \tag{4.21}$$

The radial part of the ∇^2 operator is

$$\frac{1}{r^2}\frac{\partial}{\partial r}\left(r^2\frac{\partial}{\partial r}\right)$$

so equation (2.2) reduces to the one-dimensional equation

$$-\frac{\hbar^2}{2mr^2}\frac{d}{dr}\left(r^2\frac{du}{dr}\right) - \frac{e^2 u}{4\pi\varepsilon_0 r} = Eu \tag{4.22}$$

where ∂ may be replaced by d as u is a function of r only. To avoid continuing to write the constants, we define

$$\rho = \alpha r, \quad \alpha^2 = -8mE/\hbar^2, \quad \lambda = 2me^2/4\pi\varepsilon_0\alpha\hbar^2 \tag{4.23}$$

whereupon equation (4.22) simplifies to

$$\frac{1}{\rho^2}\frac{d}{d\rho}\left(\rho^2\frac{du}{d\rho}\right) + \left(\frac{\lambda}{\rho} - \frac{1}{4}\right)u = 0 \tag{4.24}$$

As usual we expect an exponential decline for large r, and indeed $e^{-\rho/2}$ satisfies equation (4.24) as $\rho \to \infty$. Hence, as in the last section, we guess solutions of equation (4.24) will have the form $F(\rho)e^{-\rho/2}$, where F is a polynomial. Substitution into equation (4.24) shows that F must satisfy the equation

$$\frac{d^2F}{d\rho^2} + \left(\frac{2}{\rho} - 1\right)\frac{dF}{d\rho} + \left(\frac{\lambda - 1}{\rho}\right)F = 0 \tag{4.25}$$

Suppose the leading term in F is ρ^k. Substitution into the left-hand side of equation (4.25) yields the expression

$$k(k-1)\rho^{k-2} + \left(\frac{2}{\rho}-1\right)k\rho^{k-1} + \left(\frac{\lambda-1}{\rho}\right)\rho^k \qquad (4.26)$$

The leading power is ρ^{k-1}, which has coefficient $\lambda - k - 1$. This term cannot cancel against lower-order terms in the polynomial as these all have power ρ^{k-2} or less, so to satisfy equation (4.25) we must have $\lambda = k + 1$, $k = 0, 1, 2, \ldots$ or, more conventionally,

$$\lambda = n, \quad n = 1, 2, 3, \ldots \qquad (4.27)$$

Noting equation (4.23) we see that the energy levels are given by

$$E = E_n = -me^4/2(4\pi\varepsilon_0)^2\hbar^2n^2 \qquad (4.28)$$

(E_n is negative because it is conventional to make $V = 0$ at $r = \infty$.) Although this result has been derived neglecting the angular motion, we shall see that, because of degeneracy, no new energy levels are introduced by consideration of the angular variables and that equation (4.28) yields the complete set of (non-relativistic) hydrogenic energy levels. Note that this result is of precisely the form considered by Bohr in his quantization of the atom (equation (1.5)).

There are an infinite number of energy levels which get closer and closer together as $n \to \infty$. (For $E > 0$ the electron is unbound and the levels form a continuum.) This rapid ($1/n^2$) convergence of levels is due to the rapid widening of the Coulomb potential well as $V \to 0$. The lowest energy level ($n = 1$, ground state) is $-me^4/2(4\pi\varepsilon_0)^2\hbar^2$, which has the value $-13.6\,\mathrm{eV}$. This is the *ionization energy* of the hydrogen atom because it is the minimum energy needed to raise the atom from its (normal) ground state to an unbound state. Experiment confirms this calculated value.

The ground state wave function u_0 is easily deduced. For $n = 1$ ($k = 0$) the polynomial F reduces to a constant, so the solution of equation (4.25) is just $e^{-\rho/2}$. Normalizing

$$4\pi \int_0^\infty |u_0|^2 r^2 \, \mathrm{d}r = 1 \qquad (4.29)$$

where we have the volume element $\mathrm{d}\tau = 4\pi r^2 \, \mathrm{d}r$ in this spherically symmetric case, we find

$$u_0 = (\pi a_0^3)^{-1/2} e^{-r/a_0}, \quad a_0 = 4\pi\varepsilon_0\hbar^2/me^2$$

The probability of finding the electron in the volume $\mathrm{d}\tau$ is $|u_0|^2 \, \mathrm{d}\tau$,

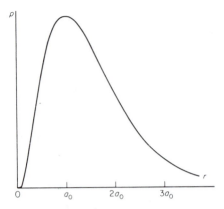

Fig. 4.2 Radial position probability density for the electron in the ground state of the hydrogen atom. The most probable position is a_0, and the average position is $3a_0/2$. The probability declines exponentially at large r.

so the probability of the electron being in the spherical shell between r and $r + dr$ is $4\pi r^2 |u_0|^2 \, dr$. The radial probability density $4\pi r^2 |u_0|^2$ is plotted in Fig. 4.2. It has a maximum at $r = a_0$. The constant a_0 is therefore the most probable radial distance of the electron from the nucleus and is known as the Bohr radius because it was originally introduced by Bohr in his atomic model. The expectation value of r for the ground state is

$$\langle r \rangle = 4\pi \int_0^\infty u_0^* r u_0 r^2 \, dr$$

$$= \frac{4}{a_0^3} \int_0^\infty r^3 e^{-2r/a_0} \, dr = \frac{3a_0}{2} \tag{4.30}$$

The fact that the electron is not fixed to a specific value of r but can be found over a range of radial distances is a significant feature of quantum mechanics. Indeed there is a small but finite probability that the electron will be found inside the nucleus itself (size $\approx 10^{-15}$ m) where it could participate in a nuclear reaction. This phenomenon is observed in some heavy atoms, where the capture of an atomic electron by the nucleus (called K capture) results in a transmutation of a proton into a neutron.

Perhaps of greater significance is the expectation value of $1/r$

$$\langle 1/r \rangle = 1/a_0 \tag{4.31}$$

Evidently the potential energy has the expectation value $-e^2/4\pi\varepsilon_0 a_0$, as in the Bohr theory.

4.3 FREE PARTICLES

If $V = 0$ we have a free particle and equation (2.7) reduces to

$$-\frac{\hbar^2}{2m}\frac{d^2u}{dx^2} = Eu \tag{4.32}$$

with solutions

$$u \propto e^{ikx}, \quad -\infty < k < \infty$$

where $|k| = (2mE/\hbar^2)^{1/2}$. Thus

$$\psi(x,t) \propto \exp[i(kx - \hbar k^2 t/2m)] = \exp[i(px - Et)/\hbar] \tag{4.33}$$

using equation (1.7). Apart from the requirement $E > 0$ there is no restriction on the value of the energy. The energy levels form a continuum.

These solutions correspond to waves which travel to the right $(k > 0)$ or left $(k < 0)$ for all time, i.e. they are infinite waves. This reflects the fact that precisely fixing k (and hence p) implies, via the Heisenberg uncertainty principle, that the location of the particle is completely indeterminate – it is equally likely to be found anywhere along the x axis. Indeed it follows from equation (4.33) that $|\psi|^2$ is constant in this case, a property which leads to a technical snag because the normalization integral diverges

$$\int_{-\infty}^{\infty} |\psi|^2 \, dx = \infty \tag{4.34}$$

The trouble arises from the over-idealization implied in the solutions (equation (4.33)). In a practical experiment the position of the particle would not be *completely* indeterminate. We would normally assume it at least to be located in the laboratory, for example. This means we would be dealing with a very long but finite wave *packet* rather than a pure oscillatory wave of unchanging amplitude. Such a packet would imply a wave function composed of a superposition of waves of slightly different momentum p. It would be inconvenient to construct wave-packet solutions in all practical calculations, and two easier methods of side-stepping the normalization problem are frequently employed. One makes use of the Dirac δ function, which is discussed in Appendix C. The other simply assumes that the particle inhabits not unbounded space but a closed and finite universe (which might actually be the case in the real universe). In one dimension this arcane-sounding criterion is simply the requirement that the particle be allowed to move freely on a circle of large circumference L rather than on an infinite line. (This is topologically equivalent to identifying points x and $x + L$.) Then equation (4.34) is replaced by

$$\int_0^L |\psi|^2 \, \mathrm{d}x = 1$$

whereupon we find

$$\psi(x, t) = L^{-1/2} \mathrm{e}^{\mathrm{i}(kx - Et/\hbar)} \qquad (4.35)$$

are correctly normalized wave functions. Because the range of x is now finite the particle is effectively trapped in a box of size L, but a box without walls. There is thus no impediment to the existence of running waves (running around the circle), unlike the case of the square well in which the vanishing of ψ at the walls forces the system to adopt standing-wave configurations.

Recalling that ψ must be a single-valued function of position, we require

$$\psi(x + L, t) = \psi(x, t)$$

so

$$\mathrm{e}^{\mathrm{i}kL} = 1$$

or

$$k = \frac{2\pi n}{L}, \quad n = 0, \pm 1, \pm 2, \pm 3, \dots \qquad (4.36)$$

The allowed momenta $\hbar k$ and energies $\hbar^2 k^2 / 2m$ are therefore discrete, but the levels are very closely spaced for large L and in practical calculations can be made indistinguishable from a continuum by letting $L \to \infty$ at the end of the calculation.

This approach is readily extended to three dimensions. We then have the plane-wave solutions

$$\psi(\mathbf{r}, t) = L^{-3/2} \mathrm{e}^{\mathrm{i}(\mathbf{k} \cdot \mathbf{r} - Et/\hbar)} \qquad (4.37)$$

where $\mathbf{k}$ is the direction of wave propagation (perpendicular to the plane of the wave front).

One now obtains sensible answers for the probability density which is simply $P(r, t) = L^{-3}$. This is independent of r and t and shows the particle is equally likely to be anywhere in the 'universe' (box) with unchanging probability.

4.4 SCATTERING FROM A POTENTIAL STEP

We may use the free-particle wave functions to describe an important physical process – the scattering of a steady beam of particles from a barrier. In section 1.6 an expression for the particle flux was derived. Substituting the wave function (equation (4.37)) into equation (1.18) yields

$$j(r, t) = \hbar k/mL^3 = p/mL^3 \qquad (4.38)$$

representing a current moving in the k direction, i.e. the direction of the momentum vector p, with the classical speed p/m, having a density of one particle per volume L^3. In one dimension $j(x, t) = p/mL$. To treat a steady flux of many independent particles we could either

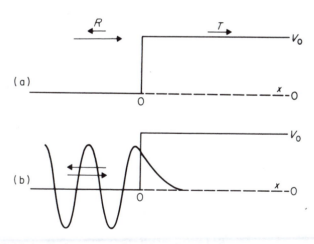

Fig. 4.3 (a) A steady stream of particles from the left with energy $E > V_0$ encounters a potential step. A fraction T of the particles continues towards the right with diminished momentum, the remaining fraction R being reflected leftwards, even though classically all the particles possess sufficient energy to surmount the barrier. (b) When $E < V_0$ all the particles are reflected. The wave function, however, penetrates a short way into the classically inaccessible region $x > 0$. This phenomenon is akin to the 'skin depth' effect for a classical electromagnetic wave reflecting from a conducting surface.

regard the particles as spaced out by L or multiply j by a numerical factor.

Interesting phenomena occur if this current impinges on an obstacle. To take a simple one-dimensional example, consider a right-moving current which encounters the finite potential step (Fig. 4.3)

$$V = 0 \qquad x < 0$$
$$= V_0 > 0 \qquad x > 0$$

The step might be expected to slow up the particles if $E > V_0$ and to reflect them back if $E < V_0$. Let us see. First suppose $E > V_0$. To the left of the step we have $\psi = u(x)e^{-iEt/\hbar}$ where

$$u(x) = L^{-1/2}(Ae^{ikx} + Be^{-ikx}) \qquad (4.39)$$

A, B and k are constants (k real and positive), A being the relative amplitude of the right-moving waves and B representing any reflected (left-moving) waves.

In the region $x > 0$, equation (2.7) reduces to

$$\frac{d^2u}{dx^2} = -\frac{2m}{\hbar^2}(E - V_0)u$$

with solutions

$$L^{-1/2}(Ce^{ik'x} + De^{-ik'x}) \qquad (4.40)$$

where $k' = [2m(E - V_0)/\hbar^2]^{1/2}$. If there is no incoming flux from right to left, i.e. coming in from $x = +\infty$, we must put $D = 0$. Then matching the two solutions, equations (4.39) and (4.40), by ensuring that the values and their derivatives are continuous at $x = 0$ yields

$$A + B = C,$$
$$k(A - B) = k'C$$

whence

and
$$\left. \begin{array}{l} B/A = (k - k')/(k + k') \\ C/A = 2k/(k + k') \end{array} \right\} \qquad (4.41)$$

The terms $L^{-1/2}Ae^{ikx}$, $L^{-1/2}Be^{-ikx}$ and $L^{-1/2}Ce^{ik'x}$ represent current densities $j = |A|^2\hbar k/mL$, $-|B|^2\hbar k/mL$ and $|C|^2\hbar k'/mL$ respectively. Thus $|B|^2/|A|^2$ represents the fraction of the flux which is reflected leftwards from the step, while $k'|C|^2/k|A|^2$ is the transmitted fraction. As required, equations (4.41) satisfy $|B/A|^2 + (k'/k)^2|C/A|^2 = 1$, so no flux disappears. We have

$$R \equiv \text{reflected fraction} = (k - k')^2/(k + k')^2 \quad\left.\right\}$$
$$T \equiv \text{transmitted fraction} = 4kk'/(k + k')^2 \quad \tag{4.42}$$

A number of interesting things are already apparent. First, even though $E > V_0$, $T < 1$, i.e. some particles bounce back. This is a purely quantum phenomenon. (For $E \gg V_0$ the classical limit is approached; one finds $k' \approx k$, so $R \approx 0$, $T \approx 1$.) Those particles that get past the barrier have $k' < k$, so they are reduced in momentum, as expected. As $E \to V_0$, $k' \to 0$ and $T \to 0$. More and more flux gets reflected.

For $E < V_0$, k' is pure imaginary and we see from equations (4.41) that $|B|^2/|A|^2 = 1$. Thus $R = 1$, $T = 0$, and *all* the particles are reflected, as in the classical case, Nevertheless the wave function does not vanish in the region $x > 0$. It is apparent from equation (4.40) that u falls off exponentially in this region: $u \propto e^{-\kappa x}$, where $k' = i\kappa$, and $D = 0$ to remove the unacceptable exponentially growing solutions. There is thus a finite probability of a particle penetrating a short way ($\approx \kappa^{-1}$) beyond the step, even though it has insufficient energy to surmount it.

By assuming $V_0 < 0$ one can consider what happens when a flux of particles encounters a step *down*. Classically this would accelerate the particles, and indeed $k' > k$ indicating a boost to the momentum. We may still use equations (4.42), from which we find that the step down also scatters some particles back even though classically it attracts them forwards. In fact for $E \ll V_0$ (low-energy incident beam) $k' \gg k$ and $R \sim 1$, $T \sim 0$. There is nearly total *reflection*. This can be important in slow-electron or neutron scattering.

4.5 TUNNELLING

The rapidly declining tail on a wave function which penetrates into a classically forbidden region of a barrier could, if the barrier were of finite thickness, emerge from the remote side and become wave-like again, representing the tunnelling of particles through the barrier.

Consider the square-hill potential (Fig. 4.4)

$$\begin{aligned} V &= 0 & x &< 0 \\ &= V_0 > 0 & 0 &< x < a \\ &= 0 & a &< x \end{aligned} \quad\left.\right\} \tag{4.4}$$

Fig. 4.4 A steady stream of particles from the left encounters a 'square-hill' potential barrier with energy $E < V_0$. In fact what is plotted is the real part of the wave function indicated by equations (4.44)–(4.46) with A chosen to be equal to unity and $ka = \kappa a = 3$ which gives a convenient scale. The choice of $k = \kappa$ corresponds to $E = \frac{1}{2} V_0$. (There is no particular significance in plotting the real rather than the imaginary part of the wave function, or indeed in plotting the real or imaginary parts of $u(x)$ rather than $\psi(x, t)$ at arbitrary time. The intention is merely to give a visual insight into the phenomenon of tunnelling.) Quantum penetration of the barrier allows the wave to emerge weakened on the remote side of the barrier, representing a finite probability that a given particle will quantum mechanically 'tunnel' through the barrier. There will thus be a transmitted fraction T and a reflected fraction R. The wave function shown decays rapidly inside the barrier so the tunnel effect is small except for low thin barriers.

Wave function solutions in these three respective regions contain the factors

$$A e^{ikx} + B e^{-ikx} \qquad (4.44)$$

$$C e^{\kappa x} + D e^{-\kappa x} \qquad (4.45)$$

$$E e^{ikx} \qquad (4.46)$$

where k and κ are defined as in section 4.4.

This time we cannot assume $D = 0$ as solution (4.45) is no longer valid out of $x = \infty$. There is no e^{-ikx} term in expression [4.46] as we assume there is no incident wave from the right. These functions and their derivatives must be matched at $x = 0$ and $x = a$. This gives four equations for the five constants A to E; normalization fixes the remaining one. We are, however, only interested in the reflection and transmission coefficients, which in this case depend on the ratios $|B|^2/|A|^2$ and $|E|^2/|A|^2$ respectively. Simple algebra gives

$$R = \left[1 + \frac{4E(V_0 - E)}{V_0^2 \sinh^2 \kappa a} \right]^{-1} \tag{4.47}$$

$$T = \left[1 + \frac{V_0^2 \sinh^2 \kappa a}{4E(V_0 - E)} \right]^{-1} \tag{4.48}$$

When the hill is thick and high $\kappa a \gg 1$ and T falls away sharply:

$$T \sim 16E(V_0 - E)e^{-2\kappa a}/V_0^2 \tag{4.49}$$

On the other hand, for $\kappa a \ll 1$,

$$T \sim 1 - (ma^2 V_0^2/2E\hbar^2) \tag{4.50}$$

and barrier penetration becomes very efficient. The ability of quantum particles to tunnel efficiently through thin barriers is exploited in a number of electronic devices. It also provides an explanation for how alpha particles escape from the nuclei of some heavy elements.

Finally, if $E > V_0$ the hill is classically surmountable but quantum effects can still be important. Putting $k' = i\kappa$ changes $\sinh \kappa a$ to $-i \sin k'a$ in equations (4.47) and (4.48). Thus when $k'a = \pi$, 2π, $3\pi, \ldots, R = 0$, $T = 1$ and there is perfect transmission; this is a sort of resonance effect. Otherwise there is some reflection, but $R \to 0$ as the classical limit ($E \gg V_0$) is approached.

4.6 MANY-PARTICLE WAVE FUNCTIONS

Until now most of the problems treated have concerned a single particle moving in a given background potential. Many of the more interesting aspects of quantum physics, however, concern multi-particle systems.

If we have a collection of N particles which do not interact and are distinguishable from each other, then each particle moves independently of the others and we can use the usual principle of multiplying probabilities to obtain a total probability. In wave mechanics this amounts to treating the total wave function Ψ as a simple product of individual one-particle wave functions

$$\Psi(r_1, r_2, \ldots, r_N, t) = \psi_1(r_1, t)\psi_2(r_2, t) \cdots \psi_N(r_N, t) \qquad (4.51)$$

where r_i is the position of particle i, etc. Then

$$|\Psi|^2 \, d\tau_1 \, d\tau_2 \cdots d\tau_N = |\psi_1|^2 \, d\tau_1 \cdot |\psi_2|^2 \, d\tau_2 \cdots |\psi_N|^2 \, d\tau_N \qquad (4.52)$$

is the probability that particle 1 is in volume $d\tau_1$, particle 2 in volume $d\tau_2$, etc. We used this elementary assumption implicitly in section 4.4 when treating the scattering of a flux of particles from a potential barrier.

Equation (2.2) is replaced by

$$\sum_{i=1}^{N} \left[-\frac{\hbar^2}{2m_i} \nabla_i^2 + V_i(r_i) \right] \Psi = E\Psi \qquad (4.53)$$

where ∇_i^2 acts on variable r_i. This separates into N equations of the form

$$\left(-\frac{\hbar^2}{2m_i} \nabla_i^2 + V_i \right) \psi_i = E_i \psi_i \qquad (4.54)$$

where

$$E = \sum_{i=1}^{N} E_i$$

so that the energies of the stationary states are, unremarkably, the sums of the energies of each individual particle.

All this changes, however, if the particles cannot be distinguished from each other. This is certainly the case for electrons taken together; one electron looks exactly like another. The same is true of collections of protons or neutrons or photons. When it comes to taking probability densities $|\Psi|^2$ we must ensure that we get the same result even if the positions of identical particles are permuted. Thus in a two-particle system if you cannot tell particle 1 from 2

$$|\Psi(r_1, r_2)|^2 = |\Psi(r_2, r_1)|^2 \qquad (4.55)$$

so either

$$\Psi(r_1, r_2) = \Psi(r_2, r_1) \qquad (4.56)$$

or

$$\Psi(r_1, r_2) = -\Psi(r_2, r_1) \qquad (4.57)$$

Neither condition is met if we use the form of equation (4.51). Instead we must take

$$\Psi(r_1, r_2) = \frac{1}{\sqrt{2}} [\psi_1(r_1)\psi_2(r_2) + \psi_2(r_1)\psi_1(r_2)] \qquad (4.58)$$

or

$$\Psi(r_1, r_2) = \frac{1}{\sqrt{2}} [\psi_1(r_1)\psi_2(r_2) - \psi_2(r_1)\psi_1(r_2)] \qquad (4.59)$$

for equations (4.56) and (4.57) respectively. The factor $1/\sqrt{2}$ is for normalization. The $+$ $(-)$ case is called symmetric (antisymmetric). Which do we choose?

In the antisymmetric case we note that Ψ vanishes when $\psi_1 = \psi_2$, i.e. the particles are in the same state. Therefore if the particles are described by an antisymmetric wave function there is zero probability that they will coincide, which means, crudely speaking, that they tend to keep away from each other. This does not happen in the symmetric case. The *ad hoc* rule that no two particles can have the same wave function ψ, i.e. occupy the same quantum state, is known as the *Pauli exclusion principle*, and was introduced originally by Wolfgang Pauli to explain why, for example, in many-electron atoms the electrons stack up in higher and higher energy levels rather than collapse together into the ground state. A study of nuclear structure reveals that protons and neutrons also obey the Pauli exclusion principle. It can be shown that this important exclusive property is related to the existence of intrinsic spin for the particles involved (section 7.4). For such particles, then, it is necessary to use *antisymmetric* wave functions.

5

Scattering in three dimensions

Although the treatment of the behaviour of beams of particles in collision with one-dimensional potential obstacles of various kinds (steps, barriers, wells, etc.) illuminates a number of interesting features, it would be a pity to stop there. After all, it would leave out all the interesting physics associated with colliding particles. As a way into that area we consider the basic problem of the scattering of a beam of particles by an immovable scattering centre which exerts a force on the moving particles depending only on the separation r. An example of this is the classic Rutherford experiment in which α-particles were scattered during passage through a thin film target containing atomic scattering centres; more broadly, most of the particle physics and ion implantation physics done in accelerators at CERN and elsewhere can be seen as resting on a proper understanding of scattering.

5.1 PARTIAL WAVE ANALYSIS

We begin by considering an idealized experimental arrangement as illustrated in Fig. 5.1.

In order to simplify the geometry, consider a parallel flux (i.e. an infinitely broad beam) of particles moving parallel to a line of incidence through the immovable scattering centre. We shall see that the beam may be represented as a sum of 'partial waves', each of which is characterized by a phase. It proves possible to describe the whole scattering process in terms of these partial waves; hence the expression 'partial wave analysis'. The incident particle current density has units of particles per unit area per second. The particles in the beam are all supposed to have the same speed v before they encounter the

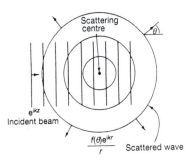

Fig. 5.1 This illustrates an incident beam, represented as a plane wave, approaching a scattering centre. Some of the incident beam is scattered and this is shown as an outgoing spherical wave, whose amplitude depends on the angle θ through the scattering length $f(\theta)$. The unscattered part of the beam continues forward as a plane wave.

scattering centre; also they are supposed to be sufficiently far apart from one another in the beam that particle–particle interaction can be safely ignored. After its encounter with the target, every particle will eventually be found to be moving away at some angle θ with respect to the line of incidence as indicated in the figure. Of course those particles whose trajectories are remote from the scattering centre will carry on in an almost straight line, that is, with θ close to zero. However, those that pass nearby become deflected by a finite θ and an observer will see free particles flying outwards from the scattering centre.

Let us first be clear about the physical quantities we wish to calculate. The concept of *scattering cross-section* is very useful in dealing with the interaction of a beam of particles with a scattering centre. When a beam with a particle flux of J particles per square metre per second is incident on a scattering centre, particles will generally be found moving away from the centre in all directions. Let dN be the number of particles scattered per second into an element of solid angle $d\Omega$ about the direction described by the polar angles θ and ϕ. We expect dN to be proportional to the incident flux J and to the size of the solid angle $d\Omega$. These notions are summarized by

$$d\sigma = dN/J = \sigma(\theta, \phi)\, d\Omega \qquad (5.1)$$

which defines $d\sigma$ (as dN/J) and $\sigma(\theta, \phi)$ (as $(1/J)\, dN/d\Omega$), both of which have the dimensions of area. In this exposition we shall deal exclusively with central potentials (potentials which depend only upon distance r from a scattering centre), and it is therefore necessary to use only one angle, θ, to characterize the direction of the trajectory after collision; $\sigma(\theta)$ will suffice in place of $\sigma(\theta, \phi)$. $\sigma(\theta)$ is known as the *differential cross-section*. The *total cross-section* σ is by definition obtainable by an integration of $\sigma(\theta)$ over all solid angles:

$$\sigma = \oint \sigma(\theta)\,d\Omega \tag{5.2}$$

We assume here that the scattering centre is fixed and time independent. Recoil could readily be included by using the centre of mass frame and replacing the mass of the incident particle by the reduced mass. In the case that the target particle is heavy, however, recoil corrections will be small.

We study the scattering problem by solving the time-independent Schrödinger equation. Conservation of energy implies that, after the scattering is complete, the freely-moving deflected particles will have the same energy and de Broglie wavelength as the incident particles. However, during the scattering process, the particles are accelerated or decelerated by the scattering potential. Thus, in the vicinity of the target the de Broglie wavelength will be altered. We therefore expect that the particle will emerge from the encounter with a shift δ in the de Broglie phase.

Because we are interested in the form of the wave function only at large r after the scattering is complete, the picture we have is that there is an incident beam which can be represented as a plane wave $\exp(ikz)$ and a scattered wave which, provided that the observer is sufficiently distant from the scattering centre, may be written as a spherical wave $(1/r)f(\theta)\exp(ikr)$. So we have

$$u(r, \theta) = \exp(ikz) + \frac{f(\theta)\exp(ikr)}{r} \tag{5.3}$$

where $f(\theta)$ is known as the *scattering amplitude*; it has the dimensions of length. It is straightforward to take the spherical wave $(1/r)f(\theta)\exp(ikr)$ and to use the operator given in equation (1.18) to derive the probability current density. Question (5.1) indicates how to do this piece of algebra and the result is that

$$d\sigma = |f(\theta)|^2\,d\Omega \tag{5.4}$$

The strategy for finding $f(\theta)$ is to study the solutions to the time-independent Schrödinger equation in the region $r \to \infty$, using the appropriate spherical polar coordinates. The full solutions as functions of r, θ, and ϕ will be discussed in Chapter 7, but for present purposes we need only remark that the wave function can be factorized as

$$u_l(r, \theta) = R(r)P_l(\cos \theta) \tag{5.5}$$

where l is a quantum number, $P_l(\cos \theta)$ is well-known (Appendix B) as the Legendre polynomial of order l, and $R(r)$ must satisfy

$$\left[-\frac{\hbar^2}{2\mu}\frac{1}{r^2}\frac{\mathrm{d}}{\mathrm{d}r}\left(r^2\frac{\mathrm{d}}{\mathrm{d}r}\right) + \frac{\hbar^2 l(l+1)}{2\mu r^2} + V(r) \right]R(r) = ER(r) \tag{5.6}$$

If we now define k and $\chi(r)$ so that $R = \chi/r$ and $E = \hbar^2 k^2/2\mu$, equation (5.6) becomes

$$\frac{\mathrm{d}^2\chi}{\mathrm{d}r^2} + \left[k^2 - \frac{l(l+1)}{r^2} - \frac{2\mu V(r)}{\hbar^2} \right]\chi = 0 \tag{5.7}$$

If $V(r)$ falls to zero faster than $1/r$ as $r \to \infty$, then at large r, equation (5.7) reduces to the simpler form

$$\frac{\mathrm{d}^2\chi}{\mathrm{d}r^2} + k^2\chi = 0 \tag{5.8}$$

Unfortunately, the requirement on the form of $V(r)$ strictly rules out the case of the Coulomb potential, for which other methods must be used. In many applications, however, the Coulomb potential is modified ('screened') in such a way as to satisfy the requirement. The general solution to equation (5.8) is

$$\chi = B\exp(+\mathrm{i}kr) + A\exp(-\mathrm{i}kr) \tag{5.9}$$

which, with equation (5.5), leads to

$$u_{kl} = \frac{1}{r}[B\exp(+\mathrm{i}kr) + A\exp(-\mathrm{i}kr)]P_l(\cos \theta) \tag{5.10}$$

The first exponential term relates to the outgoing flux and the second to the incoming flux and since the potential does not either create or annihilate particles there is a requirement that these be equal. It follows that $|A|^2 = |B|^2$ so that B may differ from A only by a phase factor. This statement is true independently of the scattering potential, but the magnitude of the phase factor is governed by the form of that potential. An alternative way of writing equation (5.10), including the phase shift δ_l, is as

$$u_{kl} = \frac{A_l}{r}\{\exp(-ikr) - \exp[+i(kr - l\pi + 2\delta_l)]\}P_l(\cos\theta) \quad (5.11)$$

The allowed energies corresponding to this wave function are independent of the quantum number l, so any linear combination of the functions u_{kl} that share the same value of k is also an acceptable function. Such a combination could be written as

$$\sum_l \frac{A_l}{r}\{\exp(-ikr) - \exp[+i(kr - l\pi + 2\delta_l)]\}P_l(\cos\vartheta) \quad (5.12)$$

In order to relate these spherical asymptotic wave functions to the form of equation (5.3) in which the incident beam is expressed as a plane wave, we use the expansion (cf. *A. Treatise on the Theory of Bessel Functions* by G. N. Watson, published by Cambridge University Press in 1958)

$$\exp(ikz) = \sum_l \left[\frac{(2l+1)i^{(2l+1)}}{kr}\right]\{\exp(-ikr)$$
$$- \exp[+i(kr - l\pi + 2\delta_l)]\}\tfrac{1}{2}P_l(\cos\theta) \quad (5.13)$$

Now finally, a little finesse produces the desired result. We expect that the total wave function, equation (5.12), minus the wave function representing the incident beam, equation (5.13), will give the scattered beam. However, the scattered beam cannot be allowed to contain any factors $\exp(-ikr)$ which would represent *incoming* particles. Physically there cannot be any of these, so all those terms must disappear, and this allows us to determine the coefficients A_l:

$$A_l = \tfrac{1}{2}[4\pi(2l+1)]^{1/2}i^{2l+1}k^{-1} \quad (5.14)$$

All that remains is algebra. Taking equation (5.12) (the total wave function), subtracting equation (5.13) (the incident wave function), and dividing by $(1/r)\exp(ikr)$ gives the result $f(\theta)$ from which the differential cross-section can be derived:

$$f(\theta) = (1/k)\sum_{l=0}^{\infty}(2l+1)\exp(i\delta_l)\sin\delta_l P_l(\cos\theta) \quad (5.15)$$

Thus the scattering problem has been reduced to that of finding the phase shifts δ_l for each of the constituent partial waves (this is why the method is described as partial wave analysis). Note that if all the δ_l were zero then $f(\theta)$ would be zero which would signify no scattering at all. If we knew all the δ_l, then equation (5.4) would give the differential cross-section $d\sigma$. Integration over all solid angles then gives (although we only quote the result of the calculation here) the total cross-section σ:

$$\sigma = \frac{4\pi}{k^2} \sum_{l=0}^{\infty} (2l+1)\sin^2 \delta_l \qquad (5.16)$$

Comparison of equations (5.15) and (5.16) gives a direct relationship between the total cross-section σ and the *forward* (i.e. $\theta = 0$) *scattering amplitude* $f(0)$. Substitution of $\theta = 0$ into equation (5.15) gives

$$f(0) = (1/k) \sum_{l=0}^{\infty} (2l+1)\exp(i\delta_l)\sin \delta_l \qquad (5.17)$$

since $P_l(1) = 1$. $f(0)$ is a complex quantity. The imaginary part of it is

$$\text{Im } f(0) = (1/k) \sum_{l=0}^{\infty} (2l+1)\sin^2 \delta_l \qquad (5.18)$$

Comparing equations (5.16) and (5.18) shows that

$$\sigma = \frac{4\pi}{k} \text{Im } f(0) \qquad (5.19)$$

This neat result, which relates the total cross-section and the forward scattering amplitude, is known as the *optical theorem*.

Calculating the phase shifts δ_l for a given potential is generally very difficult. At low energies, however, when the wavelength of the incident particles is much larger than the target, only the $l = 0$ waves, the S-waves, suffer significant scattering. A semi-classical argument shows why this is so. In Fig. 5.2 a classical trajectory is shown, and the distance known as the impact parameter is labelled as b. As the particle approaches the scatterer with momentum $\hbar k$ it has angular momentum $\hbar k b$ about the scattering centre. If the range of the interaction is characterized by a distance r_0 then a particle for which $b > r_0$ will never come near enough to the scatterer to come under its influence. So we are interested only in particles (represented by partial waves) for which $\hbar k b < \hbar k r_0$, which means those whose angular momentum is less than $\hbar k r_0$. However, quantum mechanically the allowed angular momenta are $\hbar[l(l+1)]^{1/2}$ so the required criterion is

$$\hbar[l(l+1)]^{1/2} < \hbar k r_0 \qquad (5.20)$$

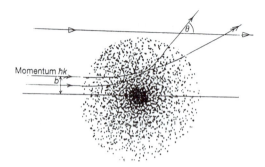

Fig. 5.2 This figure should be viewed in relation to Fig. 5.1. It shows a scattering centre which repels approaching particles and drives them into trajectories at an angle θ to the direction of approach. The angle depends on the closeness of approach.

So for sufficiently small kr_0 we need only consider the $l = 0$ particle (or partial wave). For instance, thermal neutrons of mass 1.67×10^{-27} kg moving with energy 6×10^{-21} J have $k = 4.25 \times 10^{10}$ m^{-1} and an interaction range of about 2×10^{-15} m so $kr_0 = 8.5 \times 10^{-5}$. The only value of l which satisfies equation (5.20) is thus $l = 0$.

When S-wave scattering is a good approximation, all the δ_l apart from δ_0 are zero and the expressions derived above take simpler forms. Thus for instance

$$\sigma = \frac{4\pi}{k^2} \sin^2 \delta_0 \tag{5.21}$$

5.2 ATTRACTIVE SPHERICAL WELL IN THE LIMIT OF LOW-ENERGY S-WAVE SCATTERING

A nice example of the application of S-wave scattering is that of the scattering of low energy particles of mass μ from an attractive (or repulsive) well of depth (or height) V_0 and radius a. We shall deal in some detail with the attractive case and then indicate the differences for the repulsive case. For this calculation, following on as it does from the basic theory in which the potential is taken as falling away to zero at large distances, we choose to specify the attractive well by

$$V(r) = \begin{cases} -V_0 & \text{for} \quad 0 < r < a \quad \text{(inside the well)} \\ 0 & \text{for} \quad r > a \quad \text{(outside the well)} \end{cases} \tag{5.22}$$

We treat the case $E > 0$ to exclude bound states because we are less interested in those in this context. The time-independent Schrödinger equations (TISEs) we have to solve are simplified versions of equation (5.7) with $l = 0$. Outside the well, the TISE (with $l = 0$) is

$$\frac{d^2\chi}{dr^2} + k^2\chi = 0 \tag{5.23}$$

We can select the $l = 0$ solution we want in a suitable form from equation (5.11):

$$R(r) = \frac{A_0}{r}\{\exp(-ikr) - \exp[+i(kr + 2\delta_0)]\} \tag{5.24}$$

$$= \frac{B_0\sin(kr + \delta_0)}{kr} \quad \text{(outside)} \tag{5.25}$$

where $B_0 = -2ikA_0\exp(i\delta_0)$ is used merely to simplify the appearance of equation (5.25) – it cancels out of the final answers in any case. Inside the well, the TISE is

$$\frac{d^2\chi}{dr^2} + K^2\chi = 0 \tag{5.26}$$

where $K^2 = 2\mu(E + V_0)/\hbar^2$. The acceptable solution for $R(r)$ obtained from this is

$$R(r) = \frac{C_0\sin Kr}{Kr} \quad \text{(inside)} \tag{5.27}$$

Now following the usual procedures to match the magnitudes and gradients of the wave functions (5.25) and (5.27) at the radius $r = a$ we find

$$(1/k)\tan(ka + \delta_0) = (1/K)\tan(Ka) \tag{5.28}$$

This is the equation for δ_0. In fact after a little trigonometry the solution can be written explicitly as

$$\delta_0 = \tan^{-1}[(k/K)\tan Ka] - ka \tag{5.29}$$

Finally, from equations (5.21) and (5.29),

$$\sigma = \frac{4\pi}{k^2}\sin^2\{\tan^{-1}[(k/K)\tan Ka] - ka\} \tag{5.30}$$

Clearly if the potential inside is the same as the potential outside, then $K = k$ and equation (5.29) gives $\delta_0 = 0$ as one would expect. Let us now look at the magnitude of δ_0 in a little more detail from equation (5.29). Our restriction to S-wave scattering alone is valid for small ka, ideally $ka \ll 1$ (the low energy limit). However, Ka can take any magnitude, depending simply on the depth V_0 of the well. The dashed line in Fig. 5.3 illustrates this by plotting δ_0 as a function of Ka for the case of $ka = 0.05$. We see that $\delta_0 = 0$ when $Ka = 0$, stepping up towards π as Ka rises through $\pi/2$, further steps by π also being seen whenever Ka is in the region of an odd number of $\pi/2$.

Associated with the dependence of δ_0 on Ka there is a dependence also of σ on Ka through equation (5.30). Whenever $\cot \delta_0$ is infinite, or equivalently whenever $\tan \delta_0$ is exactly zero, the cross-section will be exactly zero. These situations can be traced by examining equation (5.30) numerically and with $ka = 0.05$ as in the continuous line in Fig. 5.3 one finds that σ is zero for $Ka = 4.493, 7.725, 10.904, 14.066, 17.221$ etc. (the numerical values depend only weakly on the choice of ka provided of course that the restriction $ka \ll 1$ is respected). The separation between adjacent values is close to, but not exactly, π. What can it mean to assert that $\sigma = 0$? It is envisaged as incident particles moving past the potential well without being significantly

Fig. 5.3 This shows how δ_0(---) and σ (———) depend on Ka for the particular choice of $ka = 0.05$. The plots are of equations (5.29) and (5.30).

scattered, and there is a physical effect, the Ramsauer–Townsend effect, which is attributed to it. The effect is that electrons of energy about 0.7 eV pass through helium gas without being significantly scattered.

There is also another set of values of Ka for which σ reaches a peak of $4\pi/k^2$ as shown in the continuous line in Fig. 5.3. These occur when $\tan \delta_0 = \infty$ according to equation (5.30), and inspection of equation (5.29) shows that the crucial values correspond to $\tan Ka = \infty$ independently of the chosen value of ka. Thus, Ka takes the values of odd multiples of $\pi/2$: $Ka = 1.571, 4.712, 7.854, 10.996, 14.137, 17.279$ etc. Under these conditions the cross-section takes a peak value $4\pi/k^2$ which is greater than the geometrical cross-section πa^2 by the large factor $4/(ka)^2$. (The peaks are quite sharply defined and indeed it can be seen that all peaks except the first have a zero which is very close; thus for example there is a zero at $Ka = 17.221$ followed by a peak at $Ka = 17.279$.) Scattering at or very near a peak is called *resonant scattering*, and it can be shown to be associated with the existence of a bound state of the particle in the well, when the energy is just below the top of the well. An example is seen when low-energy neutrons are scattered from hydrogen. The observed S-wave cross-section is much greater than would be expected on the basis of the known range of the interaction and can be attributed to the existence of a bound state of the deuteron (neutron plus proton) with a very small binding energy.

5.3 REPULSIVE SPHERICAL OBSTACLE

Much of the theory of the attractive spherical well with potential $-V_0$ inside, given above, can be carried over to the repulsive spherical obstacle with $+V_0$ inside. We choose to consider the case $V_0 > E$ simply so that we can go to the limit of a hard sphere with $V_0 \rightarrow 0$. Equation (5.28) is replaced by

$$(1/k)\tan(ka + \delta_0) = (1/\kappa)\tanh(\kappa a) \tag{5.31}$$

where $\kappa^2 = 2\mu(V_0 - E)/\hbar^2$. From this we find, in place of equation (5.29),

$$\delta_0 = \tan^{-1}[(k/\kappa)\tanh \kappa a] - ka \tag{5.32}$$

This can be used to obtain σ. In the case of the totally hard sphere which has $V_0 = \infty$, it is easy to see that $\kappa = \infty$ and $\delta_0 = -ka$. In order to calculate the cross-section from equation (5.21) we need

$\sin^2 \delta_0$ with $\delta_0 = -ka$, giving $\sin^2 \delta_0 = (ka)^2$ and

$$\sigma = 4\pi a^2 \tag{5.33}$$

Thus the scattering cross-section is larger than the geometrical cross-section by a factor of 4.

We have dealt here with only the most elementary applications but clearly the methods can be used with more realistic potentials, giving insight into what hapens when particles such as neutrons, protons, atoms *etc.* collide.

6

The formal rules of quantum mechanics

Successful though a study of the wave function ψ has proved, we have yet to develop a full mechanics of microsystems that describes the measurement of general dynamical quantities. In this chapter the formal rules of quantum mechanics will be outlined.

6.1 WAVE SUPERPOSITION IN VECTOR LANGUAGE

A characteristic feature of the matter waves ψ is that, like all waves, they may be superimposed coherently:

$$\psi = c_1\psi_1 + c_2\psi_2 \qquad (6.1)$$

where c_1 and c_2 are complex numbers. For example, ψ_1 and ψ_2 could be the waves emanating from slits 1 and 2 of the Young's apparatus shown in Fig. 1.2, or they could be wave functions of a simple harmonic oscillator or a particle in a box. There is no limit to the number of such superpositions:

$$\psi = \sum_n c_n\psi_n \qquad (6.2)$$

where n may run from 1 to infinity if necessary.

It is possible to give the wave superposition a powerful vector representation. To do this we first return to the thought experiment mentioned in section 1.4, in which a particle is trapped inside a box which is then divided into two chambers, A and B, by an impenetrable membrane. The resulting wave function then consists of two non-overlapping fragments, $c_1\psi_1$ and $c_2\psi_2$, for which ψ_1 vanishes everywhere in A and ψ_2 vanishes everywhere in B. Applying the normalization condition to equation (6.1)

$$\int_{\text{box}} |\psi|^2 \, d\tau = |c_1|^2 + |c_2|^2 + 2 \operatorname{Re} c_1^* c_2 \int_{\text{box}} \psi_1^* \psi_2 \, d\tau = 1 \quad (6.3)$$

assuming ψ_1 and ψ_2 are separately correctly normalized. The probability of finding the particle in either A or B is respectively

$$\left. \begin{array}{l} \displaystyle\int_A |\psi|^2 \, d\tau = |c_1|^2 \\[2ex] \displaystyle\int_B |\psi|^2 \, d\tau = |c_2|^2 \end{array} \right\} \qquad (6.4)$$

$$|c_1|^2 + |c_2|^2 = 1 \qquad (6.5)$$

However, as the particle must be in one of the two chambers, equations (6.3) and (6.5) imply

$$\int \psi_1^* \psi_2 \, d\tau = 0 \qquad (6.6)$$

This integral is known as the 'overlap integral', for obvious reasons. That it does indeed vanish is manifest in this case, because ψ_1 and ψ_2 do not overlap, i.e. the integrand is zero everywhere: whenever ψ_1 is non-zero, ψ_2 vanishes and vice versa. That it *must* vanish is clear if we are to make sense of the probabilistic interpretation of the wave function. The particle has to be either on the right or the left. It cannot be either in both or neither.

These ideas are easily generalized to a box divided into many chambers, in which case the total wave function is described by the superposition (equation (6.2) of single-chamber wave functions, where ψ_n is the wave in chamber n. The normalization on ψ then yields

$$\int \psi_m^* \psi_n \, d\tau = 0 \qquad (n \neq m) \qquad (6.7)$$

while normalizing the individual ψ_n gives $\int \psi_n^* \psi_n \, d\tau = 1$. These two conditions may be succinctly combined to give

$$\int \psi_m^* \psi_n \, d\tau = \delta_{mn} \qquad (6.8)$$

where δ_{mn} is the Kronecker delta, defined by $\delta_{mn} = 0$ if $m \neq n$ and $\delta_{mn} = 1$ if $m = n$. (In arriving at equation (6.8) we have taken into account the fact that a linear combination of overlap integrals with arbitrary coefficients can only vanish if each individual integral vanishes.)

It is at this stage that the vector interpretation becomes apparent. A vector v in three dimensions can be expanded in terms of a set of basis vectors e_i, e_j, e_k that are *orthonormal*, i.e. mutually orthogonal and of unit length. Thus

$$v = c_i e_i + c_j e_j + c_k e_k$$

This can be readily generalized to an imaginary higher-dimensional space by considering a larger collection of orthonormal basis vectors

$$v = \sum_n c_n e_n \tag{6.9}$$

where the orthonormality condition can be written

$$e_m \cdot e_n = \delta_{mn} \tag{6.10}$$

We recognize equation (6.9) as the vector counterpart of equation (6.2), while equation (6.10) corresponds to equation (6.8). This identification is successful if we interpret the overlap integral as a scalar product of two functions ψ_m and ψ_n. This image of a wave function as being analogous to a vector is valuable to retain, and it is usual to reinforce it by using notation (Dirac notation, previously referred to the end of section 1.6) which includes the identification

$$\langle \psi_m | \psi_n \rangle \equiv \int \psi_m^* \psi_n \, d\tau \tag{6.11}$$

where the bracket $\langle | \rangle$ is used in place of the dot to denote the scalar product of two wave functions. Note that

$$\langle \psi_m | \psi_n \rangle^* = \int (\psi_m^*)^* (\psi_n)^* \, d\tau$$

$$= \int \psi_m \psi_n^* \, d\tau$$

$$= \int \psi_n^* \psi_m \, d\tau$$

$$= \langle \psi_n | \psi_m \rangle$$

We can now recognize the expansion coefficients c_n in equation (6.2) as the *components* of the 'vector' ψ in the 'direction' of the 'vector' ψ_n. From equation (6.9)

$$e_m \cdot v = \sum_n c_n e_n \cdot e_m = \sum_n c_n \delta_{nm} = c_m$$

so likewise

$$\langle \psi_m | \psi \rangle \equiv \int \psi_m^* \psi \, d\tau = c_m \tag{6.12}$$

Armed with these ideas we can construct the following mental picture of the wave function ψ. We think of ψ as a vector in some multi- (perhaps infinite-) dimensional vector space, spanned by orthonormal vectors ψ_n. The *projection* of ψ along each ψ_m is determined by the overlap integral (equation (6.12)) which yields the components c_m. Usually the ψ vector will swing around with time as ψ evolves in accordance with the Schrödinger equation. Upon making an observation to determine in which chamber the particle resides, ψ abruptly collapses onto one of the ψ_n vectors. The probability of it collapsing onto ψ_k is $|c_k|^2$, i.e. the square of the modulus of the component of ψ along ψ_k. Thus if ψ lies almost parallel to one of the ψ_n, a measurement will almost certainly find the particle in chamber n, whereas if ψ is perpendicular to some ψ_k, say, there is zero chance of finding the particle in chamber k. Alternatively we may say that if the overlap integral of ψ and ψ_k is large (small) there is a high (low) chance of finding the particle in chamber k. Finally, if a particle is known to be in chamber k, then $\psi = \psi_k$, $c_n = 0$ for all $n \neq k$; a subsequent observation will necessarily find the particle in k, i.e. $|c_k|^2 = 1$, and ψ will remain unchanged as ψ_k. No further 'collapse' occurs.

So far the vector language is simply a convenient but completely equivalent means of discussing overlap integrals of wave functions. No new physical principles are involved. New physics is injected, however, when the ideas are generalized from consideration of the *position* of the particle to other dynamical observables such as momentum, angular momentum, energy. In embarking upon this generalization, two important points must be kept in mind.

1. The orthogonality of the wave functions ψ_n is intimately associated with the *exclusive* property of measurement. The location of the electron can only be in *one* of the chambers. This either/or property is reflected in the fact that the 'vectors' ψ_n representing the alternative results of observation are mutually orthogonal

and so cannot 'overlap'. If ψ_n had a non-zero projection onto ψ_m then there would be a probability that a particle in chamber n might also be in chamber m, which is absurd.

2. A general state ψ is a superposition (equation (6.2)) of states ψ_n representing alternative outcomes of observation. Once the observation is accomplished there is an abrupt collapse of the wave function onto one of these 'basis' states. Which state, i.e. which result is obtained, is an inherently probabilistic feature, the probability being determined by the expansion coefficient, i.e. 'component', c_n.

We already have at hand the means to apply this generalization to *energy*, because E is incorporated in the time-independent Schrödinger equation. First we address point 1 above. Suppose the states ψ_n represent the stationary *energy level* states of the sort discussed in Chapters 2 and 4. Are these states orthogonal under the definition (6.7) of scalar product $\langle \psi_m | \psi_n \rangle$? In the case of the 'chamber' states associated with the location of the particle, equation (6.8) is trivially correct because the integrand itself vanishes. This is not, however, the case for the stationary states that represent energy levels. Nevertheless it is easy to show that equation (6.7) is in fact satisfied even in this case.

Consider two one-dimensional stationary states with respective wave functions and energies u_m, u_n, E_m, E_n. (The three-dimensional case follows similar lines.) The Schrödinger equation yields

$$-\frac{\hbar^2}{2m}\frac{d^2u_n(x)}{dx^2} + V(x)u_n(x) = E_n u_n(x) \qquad (6.13)$$

$$-\frac{\hbar^2}{2m}\frac{d^2u_m^*(x)}{dx^2} + V(x)u_m^*(x) = E_m u_m^*(x) \qquad (6.14)$$

Multiply equations (6.13) and (6.14) from the left by u_m^* and u_n respectively, subtract, and integrate the result. This gives

$$-\frac{\hbar^2}{2m}\int\left(u_m^*\frac{d^2u_n}{dx^2} - u_n\frac{d^2u_m^*}{dx^2}\right)dx = (E_n - E_m)\int u_m^* u_n\,dx$$

the V terms having cancelled. Integration by parts causes the left-hand integral to vanish, leaving only the boundary terms

$$-\frac{\hbar^2}{2m}\left[u_m^*\frac{du_n}{dx} - u_n\frac{du_m^*}{dx}\right]_{-\infty}^{\infty} = (E_n - E_m)\langle u_m | u_n \rangle$$

where on the right-hand side we have used Dirac notation from equation (6.11):

$$\langle u_m | u_n \rangle \equiv \int u_m^* u_n \, d\tau$$

Assuming u_m and u_n fall to zero as $x \rightarrow \pm \infty$, then

$$(E_n - E_m)\langle u_m | u_n \rangle = 0 \qquad (6.15)$$

Equation (6.15) implies two possibilities. Either $E_n = E_m$, in which case the energy levels are degenerate and a measurement of E would not lead to an either/or discrimination between states m and n, or

$$\langle u_m | u_n \rangle = 0 \qquad (6.16)$$

and the wave functions are orthogonal as required. A measurement of E then leads to *either* E_n or E_m.

A simple example of these ideas is provided by the energy levels of a particle confined to an infinite square-well potential, a problem treated in section 2.2. The stationary state wave functions for odd n are

$$\psi_n(x, t) = a^{-1/2} \cos(n\pi x/2a) \exp(-iE_n t/\hbar) \quad |x| < a$$
$$= 0 \qquad\qquad\qquad\qquad\qquad\qquad |x| > a \quad (6.17)$$

where E_n is given by equation (2.11). Orthonormality is easily checked. From equation (6.11)

$$\langle \psi_m | \psi_n \rangle = a^{-1} e^{i(E_m - E_n)t/\hbar} \int_{-a}^{a} \cos(m\pi x/2a) \cos(n\pi x/2a) \, dx$$
$$= \delta_{mn} \qquad (6.18)$$

a result which follows by noting that the integral vanishes if $m \neq n$, but equals a if $m = n$. Therefore the functions $\cos(n\pi x/2a)$ may be used as a basis of 'vectors' in terms of which an even function on the interval $-a < x < a$ may be expanded. Suppose at time $t = 0$ the wave function is $u(x)$. Then such an expansion is

$$u(x) = a^{-1/2} \sum_{n=1}^{\infty} c_n \cos(n\pi x/2a) \quad (n \text{ odd}) \qquad (6.19)$$

where according to equation (6.12)

$$c_m = a^{-1/2} \int_{-a}^{a} u(x) \cos(m\pi x/2a) \, dx$$

This will be recognized as a Fourier cosine series, with c_m being the Fourier amplitudes.

Let us consider a reasonably realistic example as illustrated in Fig. 6.1. A particle is in its ground state in an infinite square-well potential with walls at $\pm a/2$, that is, the well is of *half* the width of that discussed above. It is in its ground state with wave function

$$\psi(x,t) = (a/2)^{-1/2} \cos(\pi x/a) \exp(-i\pi^2 \hbar t/2ma^2) \qquad (6.20)$$

At $t=0$, the two walls are moved out to $\pm a$ sufficiently fast that the wave function is left standing:

$$\psi(x,0) = (a/2)^{-1/2} \cos(\pi x/a) \qquad (6.21)$$

This wave function, which was allowed for the old boundary conditions (wall separation a at $t<0$ is not acceptable for the new boundary conditions (wall separation $2a$ at $t>0$, but can be expanded in terms of the new functions given by equation (6.17) with $t=0$,

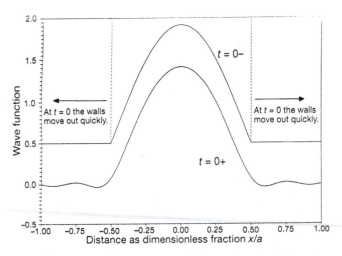

Fig. 6.1 The upper part of the figure shows the wave function before the walls are moved. Immediately after the walls are moved the wave function looks identical, but mathematically it must be represented as the sum of the eigenfunctions appropriate to the changed situation. The first six coefficients are given in the text, and the lower part of the figure shows the reconstruction including only those six. The ripple could be eliminated by taking a greater number of terms.

with the result

$$(a/2)^{-1/2}\cos(\pi x/a) = a^{-1/2}\sum_{n=1}^{\infty} c_n\cos(n\pi x/2a) \quad (n \text{ odd}) \quad (6.22)$$

The coefficients c_m can be found by multiplying both sides by $a^{-1/2}\cos(m\pi x/2a)$ and integrating between $+a/2$ and $-a/2$. The result, after some algebraic reduction, is that

$$c_m = \left(-2^{1/2}\cos\frac{m\pi}{4}\right)\frac{8}{\pi(m^2-4)} \quad (6.23)$$

The bracketed quantity has a value ± 1, the sign depending simply on m, and the first few values of c_m are $c_1 = +8/3\pi$, $c_3 = +8/5\pi$, $c_5 = +8/21\pi$, $c_7 = -8/45\pi$, $c_9 = -8/77\pi$, $c_{11} = +8/117\pi$. Figure 6.1 shows that even these first six coefficients make a convincing job of reconstructing the original wave function. A complete set would of course give a perfect reconstruction.

After the walls have been moved, for $t > 0$, the wave function has a periodic time development:

$$a^{-1/2}\sum_{n=1}^{\infty} c_n\cos(n\pi x/2a)\exp(-in^2E_1t/\hbar) \quad (6.24)$$

where $E_1 = \pi^2\hbar^2/8ma^2$ is the ground state energy. (It is not difficult to show (Exercise 6.1) that the time dependence of the wave function is periodic with periodic time h/E_1.)

If the system is initially in a superposition (equation (6.2)) of many possible stationary energy-level states ψ_n, then, after measurement, we must find a *particular* energy, say E_k, in which case the state ψ abruptly changes to state k with wave function ψ_k.

We now extend the probability interpretation of the components c_n to these energy states and postulate that if ψ is initially given by equation (6.2) then *on measurement of the energy there is a probability* $|c_k|^2$ *that the value E_k will be obtained*.

Armed with this crucial postulate we can immediately compute a formula for the average or *expectation* value of the energy E. If E_n has a probability $|c_n|^2$ then

$$\langle E \rangle = \sum_n E_n|c_n|^2 \quad (6.25)$$

A more useful form of equation (6.25) is (returning to three dimensions)

$$\langle E \rangle = \int \psi^* \left(-\frac{\hbar^2}{2m} \nabla^2 + V \right) \psi \, d\tau \tag{6.26}$$

The equivalence of equations (6.25) and (6.26) follows using equation (6.2)

$$\langle E \rangle = \int \sum_m c_m^* \psi_m^* \left(-\frac{\hbar^2}{2m} \nabla^2 + V \right) \sum_n c_n \psi_n \, d\tau$$

$$= \sum_m \sum_n c_m^* c_n \int \psi_m^* \left(-\frac{\hbar^2}{2m} \nabla^2 + V \right) \psi_n \, d\tau \tag{6.27}$$

Noting that ψ_n is a stationary state, it follows from the Schrödinger equation (equation (2.2)) that

$$\left(-\frac{\hbar^2}{2m} \nabla^2 + V \right) \psi_n = E_n \psi_n$$

Thus equation (6.27) reduces to

$$\sum_m \sum_n c_m^* c_n E_n \langle \psi_m | \psi_n \rangle$$

$$= \sum_m \sum_n c_m^* c_n E_n \delta_{mn}$$

$$= \sum_n E_n |c_n|^2$$

as claimed.

Once again, an elegant illustration is provided by the infinite square-well problem. Suppose at $t = 0$ the system is in a state with wave function

$$u(x) = a^{-1/2} \left[\frac{1}{2} \cos(\pi x/2a) + \frac{\sqrt{3}}{2} \cos(3\pi x/2a) \right] \qquad |x| < a$$

$$= 0 \qquad |x| > a$$

This is a normalized superposition of the ground state ($n = 1$) with the state $n = 3$. It is not itself a stationary state. The full time-dependent wave function will be (for $|x| < a$)

$$\psi(x, t) = a^{-1/2} \left[\frac{1}{2} e^{-iE_1 t/\hbar} \cos(\pi x/2a) + \frac{\sqrt{3}}{2} e^{-iE_3 t/\hbar} \cos(3\pi x/2a) \right] \tag{6.28}$$

where E_1 and E_3 are given by equation (2.11).

First, normalization may be checked by substituting equation (6.28) into equation (1.14):

$$\int_{-\infty}^{\infty} |\psi(x,t)|^2 \, dx = \frac{1}{4a} \int_{-a}^{a} \cos^2(\pi x/2a) \, dx$$

$$+ \frac{3}{4a} \int_{-a}^{a} \cos^2(3\pi x/2a) \, dx$$

$$= \tfrac{1}{4} + \tfrac{3}{4} = 1$$

The expectation value of the energy may be evaluated using equation (6.26). Recalling that $V = 0$ inside the well, we have

$$\langle E \rangle = -\frac{\hbar^2}{2m} \int \psi^*(x,t) \frac{\partial^2}{\partial x^2} \psi(x,t) \, dx$$

Substituting equation (6.28) into the integrand we get

$$\langle E \rangle = \frac{\hbar^2 \pi^2}{32ma^3} \left[\int_{-a}^{a} \cos^2(\pi x/2a) \, dx + 27 \int_{-a}^{a} \cos^2(3\pi x/2a) \, dx \right]$$

Note that the time dependence has dropped out. This is because the cross-terms in the integrand vanish by virtue of the orthogonality of the stationary-state wave functions, while in the remaining terms the time dependence enters as a pure phase factor which disappears as a result of the complex conjugation. Evaluating the integrals one finds $\langle E \rangle = 7\pi^2 \hbar^2/8ma^2$ (independent of time in spite of the fact that equation (6.28) is not a stationary state).

Finally, we can deduce the relative probability that, on measurement, the system will be found in either the ground state or the excited state. According to our fundamental postulate these are given by the square modulus of the amplitudes (coefficients) of the respective normalized wave functions in the superposition (equation (6.28). Thus the ground state is found with probability $\tfrac{1}{4}$, the excited state with probability $\tfrac{3}{4}$. Noting from equation (2.11) that the ground state has energy $\pi^2 \hbar^2/8ma^2$, while the state $n = 3$ has energy $9\pi^2 \hbar^2/8ma^2$, we may weight the relative probabilities with the respective energies of these states, $\tfrac{1}{4}(\pi^2 \hbar^2/8ma^2) + \tfrac{3}{4}(9\pi^2 \hbar^2/8ma^2)$, to recover $7\pi^2 \hbar^2/8ma^2$ for $\langle E \rangle$.

The association of wave functions with vectors is actually more than a mere analogy. It is possible to formulate quantum mechanics abstractly in the language of generalized vectors and to regard wave functions as simply one particular representation of the vector algebra. Other representations exist and are sometimes used, e.g. the matrix formulation where vectors are row and column matrices. Following Dirac, the quantum state vectors are usually denoted $|\rangle$, and called 'kets'. That is why we have used the notation $\langle|\rangle$ for a scalar product. The other half of this symbol, $\langle|$, is called a 'bra' vector (hence $\langle|\rangle$ is a 'bra–ket' or bracket). Ket vectors must not be confused with ordinary vectors in real three-dimensional space. In this book the notation $\langle|\rangle$ is, however, used only as a convenient shorthand and the reader need not be concerned with learning the abstract details of Dirac's formulation.

One final mathematical point should be mentioned. If a vector is to be expanded in terms of a set of orthonormal basis vectors then there must be as many basis vectors as there are dimensions in the space. Thus in ordinary three-dimensional space one needs three basis vectors e_i, e_j, e_k. Such vectors are said to form a *complete* set. In quantum mechanics one frequently deals with infinite-dimensional space and the question of completeness is mathematically delicate. Simple counting will not do; remove a few vectors from an infinite set and you still have an infinite set. This subtlety will not be pursued. We shall always assume that the wave functions that arise from the standard physical problems to be considered do form a complete set.

6.2 OPERATORS

Suppose we write the time-independent Schrödinger equation (equation (2.2)) as follows

$$\hat{H}u_n = E_n u_n \tag{6.29}$$

for the stationary state with energy E_n, where

$$\hat{H} \equiv -\frac{\hbar^2}{2m}\nabla^2 + V \tag{6.30}$$

The object $\hat{H}$ is known mathematically as an *operator*. This operator consists of two terms, the first of which is a differential operator while the second (V) is a simple multiplier. In general, operators act on functions and alter them in some way. The $\hat{\ }$ symbol (often pronounced as 'hat'), is used to distinguish an operator from an ordinary number or function, and in conversation about equation (6.30), one would say 'aitch hat' for $\hat{H}$ and 'del squared' for ∇^2.

Equation (6.29) thus states a rather special condition. It says that the function u_n, when acted upon by $\hat{H}$, shall remain unchanged except in magnitude. In vector language the vector $|u_n\rangle$ alters only in length, not direction. This rather restrictive requirement is known as an *eigenvalue* equation, the E_n being the eigenvalues, and the u_n the *eigenfunctions* (or, equivalently, eigenvectors) to which they are said to 'belong'. The quantum states associated with these eigenfunctions are known as *eigenstates* of the relevant operator.

Equation (6.29) describes physically the location of the energy levels of a stationary quantum system; the allowed energies E_n are just the eigenvalues of the operator (equation (6.30)). For that reason the object (6.30) is known as the *energy operator* or, more usually, as the *Hamiltonian* operator, which is why it is denoted by $\hat{H}$. The use of the term 'Hamiltonian' is taken from classical mechanics where, for most systems of interest, the Hamiltonian function is identical with the total energy. *Thus stationary states are energy* ($\hat{H}$) *eigenstates.* We may now write the Schrödinger equation (1.12) in the form

$$\hat{H}\psi = i\hbar\frac{\partial\psi}{\partial t} \qquad (6.31)$$

The fact that the measured values of the energy are given by the eigenvalues of the operator $\hat{H}$ suggests that other dynamical quantities, such as momentum and angular momentum, can also be represented by operators and that their eigenvalues are the possible measured values of the associated dynamical quantity. This step cannot be proved but is one of the basic postulates of quantum mechanics, to be tested by experiment.

For this strategy to be successful we must ensure that the eigenvalues of these yet-to-be-discovered operators are real numbers, as they correspond to measurable quantities. Moreover, remembering point 1 on page 68, we also require that the corresponding eigenfunctions (eigenvectors) are orthogonal. These two conditions impose a strong restriction on the type of operator that $\hat{H}$ is allowed to be.

Writing a general eigenvalue equation as

$$\hat{A}\psi_n = \lambda_n\psi_n \qquad (6.32)$$

where now $\hat{A}$ is a general operator and ψ_n a general eigenfunction (not necessarily the Hamiltonian operator or its associated stationary-state wave functions) we have

$$\psi_m^*\hat{A}\psi_n = \lambda_n\psi_m^*\psi_n \qquad (6.33)$$

and, similarly,

$$\psi_n \hat{A}^* \psi_m^* = \lambda_m^* \psi_m^* \psi_n \qquad (6.34)$$

Subtracting equation (6.34) from equation (6.33) and integrating

$$\int \psi_m^* \hat{A} \psi_n \, d\tau - \int \psi_n \hat{A}^* \psi_m^* \, d\tau$$

$$= (\lambda_n - \lambda_m^*) \langle \psi_m | \psi_n \rangle \qquad (6.35)$$

Consider first the case $m = n$. Then on the right of equation (6.35) $\langle \psi_n | \psi_n \rangle = 1$. If the eigenvalues λ_n are all to be real, $\lambda_n - \lambda_n^* = 0$, so the right-hand side of equation (6.35) vanishes. On the other hand, if $m \neq n$, $\lambda_n - \lambda_m^* \neq 0$ (except in the case of degeneracy, $\lambda_n = \lambda_m$) but now $\langle \psi_m | \psi_n \rangle = 0$ because the eigenfunctions are required to be orthogonal. Thus in either case

$$\int \psi_m^* \hat{A} \psi_n \, d\tau = \int \psi_n \hat{A}^* \psi_m^* \, d\tau$$

Operators $\hat{A}$ for which

$$\int \phi^* (\hat{A} \psi) \, d\tau = \int \psi (\hat{A} \phi)^* \, d\tau \qquad (6.36)$$

for arbitrary functions ψ, ϕ (not necessarily eigenfunctions of $\hat{A}$ as above) belong to a special class known as *Hermitian* operators. It can be shown that the requirement that equation (6.36) be satisfied ensures that Hermitian operators always have (i) real eigenvalues and (ii) the property that eigenfunctions belonging to different eigenvalues are orthogonal. These are certainly desirable properties. The mathematics of equation (6.36) may seem rather esoteric, so we give a concrete example by asking whether either or both of the differential operators $i\,d/dx$ and d/dx are Hermitian. It turns out that the factor i makes all the difference. With the first operator, equation (6.36) becomes

$$i \int \phi^* \frac{d\psi}{dx} \, dx = \int \psi \left(i \frac{d\phi}{dx} \right)^* dx = -i \int \psi \frac{d\phi^*}{dx} \, dx \qquad (6.37)$$

and this is *generally* true because it can be rearranged as $\int d(\phi^* \psi) = 0$ which is true for all well-behaved normalizable functions. Now let us consider the second, seemingly similar, case of d/dx. We find the left and right sides of equation [6.36] are equal to

$$\int \phi^* \frac{d\psi}{dx} dx \quad \text{and} \quad \int \psi \frac{d\phi^*}{dx} dx \tag{6.38}$$

respectively, and it is easy to see that these are *not* generally equal. A simple example is to take $\phi = (1 + x)\exp(-x^2/2)$ and $\psi = (1 - x)\exp(-x^2/2)$. These are well-behaved normalizable functions in the range $-\infty$ to $+\infty$. It is straightforward to show that

$$\int_{-\infty}^{+\infty} \phi^* \frac{d\psi}{dx} dx = \int_{-\infty}^{+\infty} \phi \frac{d\psi}{dx} dx = -\pi^{1/2}/2 \tag{6.39}$$

and

$$\int_{-\infty}^{+\infty} \psi \frac{d\phi^*}{dx} dx = \int_{-\infty}^{+\infty} \psi \frac{d\phi}{dx} dx = +\pi^{1/2}/2 \tag{6.40}$$

The difference in sign of these two immediately shows that they cannot be equal, and this example is sufficient to rule out d/dx as a Hermitian operator. The same choice of functions used to test the operator $i\,d/dx$ works perfectly well of course as we know in advance from the general argument given above. So $i\,d/dx$ is Hermitian but d/dx is not.

Because of the essential properties of Hermitian operators (real eigenvalues and orthogonal eigenfunctions) it is natural to look to them to represent physical observables.

There is a close relation between Hermitian operators and Hermitian matrices. If there is a set of general functions ϕ_m, $m = 1, 2, 3, \ldots$, we can define a matrix of complex numbers as follows

$$A_{mn} \equiv \int \phi_m^* \hat{A} \phi_n \, d\tau \tag{6.41}$$

Then equation (6.36) implies $A_{mn} = A_{nm}^*$, which is the definition of an Hermitian matrix.

In Dirac notation the matrix element A_{mn} is $\langle \phi_m | \hat{A} \phi_n \rangle$. However, there is a symmetry implied by equation (6.36) in the way $\hat{A}$ acts. It may either be regarded as acting forwards on ϕ_n or, complex conjugated, backwards on ϕ_m. This means $\langle \phi_m \hat{A} | \phi_n \rangle = \langle \phi_m | \hat{A} \phi_n \rangle$ so one usually writes both equivalent expressions in symmetric form as $\langle \phi_m | \hat{A} | \phi_n \rangle$.

If the functions ϕ_m and ϕ_n are chosen to be eigenfunctions of $\hat{A}$, it follows from equation (6.32) that

$$A_{mn} = \lambda_n \int \psi_m^* \psi_n \, d\tau \equiv \lambda_n \langle \psi_m | \psi_n \rangle = \lambda_n \delta_{mn}$$

In this case the matrix A_{mn} is *diagonal*, its components being the eigen-values λ_n, which are *the possible values of the observable associated with* $\hat{A}$ which would be found on measurement. The expectation value of the observable in a general state ψ is, from an argument identical to that used to prove equation (6.26) given by

$$\langle \hat{A} \rangle = \int \psi^* \hat{A} \psi \, d\tau \tag{6.42}$$

Thus equation (6.26) is a special case of equation (6.42) with $\hat{A} = \hat{H}$, the Hamiltonian operator. In Dirac notation equation (6.42) is $\langle \psi | \hat{A} | \psi \rangle$ which motivates the abbreviated use of the brackets $\langle \rangle$. Note that if the system is in an eigenstate of $\hat{A}$, say ψ_n, then

$$\langle \hat{A} \rangle = \langle \psi_n | \hat{A} | \psi_n \rangle = \lambda_n \langle \psi_n | \psi_n \rangle = \lambda_n \tag{6.43}$$

This is expected, as we know that, when the state is an eigenstate of an operator, a measurement of its associated observable *must* yield the eigenvalue belonging to that eigenstate.

The matrices defined above can be regarded as operators which act on vectors to give other (transformed) vectors. The operator algebra and the matrix algebra are identical. Historically this matrix approach was taken by Heisenberg in parallel with Schrödinger's wave mechanics. In some cases, e.g. the discussion of angular momentum, the matrix approach is often more convenient.

6.3 MOMENTUM AND POSITION EIGENSTATES

The remaining task is to decide which Hermitian operators to choose for the various observables. Here one is assisted by Bohr's so-called *correspondence principle* which states that the relationships between various observable dynamical quantities at the quantum-operator level ought to be consistent with the classical relationships, otherwise there might be problems in taking the classical limit.

For example if the total energy E is the sum of kinetic and potential energies, $E = p^2/2m + V$. Comparison with the Hamiltonian operator, equation (6.30), then suggests the form for the *momentum operator*

$$\hat{p} = -i\hbar\nabla \quad \text{or} \quad -i\hbar\frac{\partial}{\partial x} \tag{6.44}$$

in three dimensions and one dimension respectively. The use of the minus sign will be explained below.

The eigenfunctions of $\hat{p}$, denoted by u_p, are determined by the eigenvalue equation

$$\hat{p}u_p \equiv -i\hbar\nabla u_p = pu_p \tag{6.45}$$

for eigenvalues p. (We use the symbol u rather than ψ because momentum eigenfunctions do not depend on time.) Solutions of equation (6.45) are found immediately:

$$u_p = L^{-3/2}e^{ip\cdot r/\hbar} \equiv L^{-3/2}e^{ik\cdot r} \tag{6.46}$$

using the box normalization described in section 4.3. Thus momentum eigenstates are described by plane wave eigenfunctions with the wave numbers k related to the measured values of the momentum p by de Broglie's relation $p = \hbar k$. Had a plus sign been used in equation (6.44) instead this relation would read $p = -\hbar k$, implying that the particles move in the opposite direction to the wave.

Orthonormality of the momentum eigenfunctions is easily verified. In one dimension, for example,

$$\langle u_{p'}|u_p\rangle = L^{-1}\int_0^L e^{i(k-k')x}\,dx = L^{-1}\left(\frac{e^{i(k-k')L} - 1}{i(k-k')}\right) \tag{6.47}$$

Applying the box boundary conditions (equation (4.36)), $\exp(ikL) = \exp(ik'L) = 1$ implies equation (6.47) vanishes unless $k' = k$, in which case $\langle u_{p'}|u_p\rangle = 1$.

The position observable is associated with x. This is a rather trivial operator because it acts by ordinary multiplication. Formally, one can still write $\hat{x}$ and study the eigenvalue equation

$$\hat{x}\psi_{x'} = x'\psi_{x'} \tag{6.48}$$

The eigenfunctions $\psi_{x'}$ correspond to a state in which the particle is located precisely at the point $x = x'$. Recalling that the wave function is related to the position probability density, this implies that $\psi_{x'}$ must be infinitely concentrated at $x = x'$ and vanish elsewhere. No ordinary function has this singular property, but we could consider, for example, the x axis divided into cells and ask which cell the particle inhabits. As the size of the cells decreases, so the cell wave functions approach idealized eigenfunctions $\psi_{x'}$.

A formal mathematical object that possesses these properties was introduced by Dirac and called the Dirac delta function. It is discussed further in Appendix C.

6.4 COMPATIBLE OBSERVABLES AND COMMUTATION RELATIONS

In Chapter 1 it was pointed out how the determination of a particle's position introduced an uncertainty in its momentum and vice versa. When two observables tangle with each other in this way they are called *incompatible*. However, there is no impediment to the simultaneous determination of precise values for, say, the x component of position and the y component of momentum (or position), or, in the case of free-particle states, of energy and momentum. Observables that can be independently measured at the same instant without disrupting each other are called *compatible*.

To establish a mathematical criterion for compatibility we recall that the act of measurement causes the abrupt collapse of ψ onto one of the eigenvectors of the operator associated with the observable. Suppose we have two observables, α and β, with associated operators $\hat{A}$ and $\hat{B}$ possessing eigenvectors ψ_n, ϕ_n and eigenvalues λ_n, μ_n respectively. A measurement of α will give some value, λ_k say, and throw the system into a state ψ_k. If ψ_k happens to coincide with one of the eigenvectors ϕ_n, say ϕ_k, then a measurement of β will certainly (with probability one) yield the value μ_k and the state will remain undisturbed as $\psi_k \, (= \phi_k)$. Subsequent remeasurements of α or β will always reproduce the values λ_k or μ_k and the state will remain the same. There is thus no impediment to the repeated simultaneous precise determination of both α and β. They will be compatible observables.

On the other hand, if the eigenvectors of $\hat{A}$ do not coincide with those of $\hat{B}$ the situation is dramatically different. In that case the state ψ_k, achieved as a result of the first (α) measurement, does not lie along one of the ϕ_k but will be a superposition of many ϕ_n:

$$\psi_k = \sum_n c_n \phi_n \qquad (6.49)$$

The rules of quantum mechanics then require that a measurement of β yields any particular ϕ_n with a probability $|c_n|^2$. The state resulting from the second (β) measurement might then be any one of a number of states ϕ_n and the value obtained for β might be any associated μ_n. Instead of remaining in the state ψ_k after the second measurement, the state vector (wave function) will jump yet again, this time to an eigenstate, say ϕ_m, of β.

It is now easy to see that a subsequent remeasurement of α might very well not reproduce the value λ_k found on the first occasion, for the β eigenstate ϕ_m can now be expanded

$$\phi_m = \sum_n d_n^{(m)} \psi_n \tag{6.50}$$

from which we see that the probability of getting λ_k is $|d_k^{(m)}|^2$, which is < 1. Thus with repeated alternating measurements of α and β different results are generally obtained each time around and the state vector jumps discontinuously on every measurement. Clearly, then, one cannot assign precise values to both α and β at the same time. Only the observable last measured will have a precise value.

It follows that, mathematically, the criterion for simultaneous measurability (compatibility) is that the two operators concerned should possess a common set of eigenvectors (eigenfunctions), in which case, $\hat{A}\psi_n = \lambda_n\psi_n$, $\hat{B}\psi_n = \mu_n\psi_n$. It follows that

$$\hat{B}\hat{A}\psi_n = \lambda_n\hat{B}\psi_n = \lambda_n\mu_n\psi_n \tag{6.51}$$

and

$$\hat{A}\hat{B}\psi_n = \mu_n\hat{A}\psi_n = \mu_n\lambda_n\psi_n \tag{6.52}$$

Subtracting equation (6.51) from equation (6.52) we obtain

$$(\hat{A}\hat{B} - \hat{B}\hat{A})\psi_n = 0 \tag{6.53}$$

Being valid for all ψ_n we deduce from equation (6.53) the operator equation

$$\hat{A}\hat{B} - \hat{B}\hat{A} = 0 \tag{6.54}$$

The quantity $\hat{A}\hat{B} - \hat{B}\hat{A}$ is known as the *commutator* of $\hat{A}$ and $\hat{B}$ and is abbreviated to $[\hat{A}, \hat{B}]$. Thus equation (6.54) may be written

$$[\hat{A}, \hat{B}] = 0 \tag{6.55}$$

Operators for which the commutator vanishes are said to *commute*. We conclude that *compatible observables must be represented by commuting operators*. It may be that a whole set of operators will intercommute, in which case several observables will be compatible.

If an eigenvector is common to two or more operators, it can carry two or more labels to characterize it. Thus an eigenvector of, say, energy and momentum could be labelled ψ_{Ep}. In Dirac notation one often drops the ψ altogether and just puts the eigenvalue labels in the ket vector, thus $|E, p\rangle$. Such labels are often referred to as *quantum numbers* when they are restricted to discrete values.

There ought to be a close relation between compatibility and the Heisenberg uncertainty relation, for if two observables are compatible there is no fundamental impediment to their simultaneous measurement. If the uncertainty $\Delta\alpha$ in observable α is defined precisely in the root mean square sense, $(\Delta\alpha)^2 = \langle \hat{A}^2 \rangle - \langle \hat{A} \rangle^2$, and similarly for β, then it can be proved quite generally that

$$\Delta\alpha \, \Delta\beta \geqslant \tfrac{1}{2} |\langle [\hat{A}, \hat{B}] \rangle| \qquad (6.56)$$

Thus if $\hat{A}$ and $\hat{B}$ commute, $\Delta\alpha \, \Delta\beta$ can be zero and there is no mutual uncertainty. The proof of equation (6.56) is a little too lengthy to include here. It exploits the fact that all operators representing physical variables are Hermitian, and it makes use of a mathematical theorem known as the Schwartz inequality (this inequality accounts for the inequality in equation (6.56)) which states the mathematical fact that for any two complex functions f and g the following result holds:

$$\int f^*f \, d\tau \int g^*g \, d\tau \geqslant \left| \int f^*g \, d\tau \right|^2 \qquad (6.57)$$

The possible incompatibility of observables is an essentially quantum phenomenon; it does not occur in the classical limit. Hence we expect $\hbar$ to make an appearance at this stage. Suppose, for example, α and β are one-dimensional position and momentum respectively. Then

$$[\hat{x}, \hat{p}_x]\psi = -i\hbar x \frac{\partial\psi}{\partial x} + i\hbar \frac{\partial(x\psi)}{\partial x} = i\hbar\psi \qquad (6.58)$$

As this is true for arbitrary ψ we arrive at the operator identity

$$[\hat{x}, \hat{p}_x] = i\hbar \qquad (6.59)$$

known as the *commutation relation* for $\hat{x}$ and $\hat{p}_x$. Inserting equation (6.59) in equation (6.56) gives $\Delta x \, \Delta p \geqslant \tfrac{1}{2}\hbar$ which is a precise statement of the Heisenberg uncertainty relation (equation (1.11)). Thus $\hbar$ determines the scale of the quantum fuzziness which produces the incompatibility. The operators $\hat{y}$, $\hat{p}_y$ and $\hat{z}$, $\hat{p}_z$ satisfy commutation relations similar to equation (6.59). All other commutators of $\hat{x}, \hat{y}, \hat{z}, \hat{p}_x, \hat{p}_y, \hat{p}_z$ vanish, e.g. $[\hat{x}, \hat{p}_y] = 0$.

6.5 SYMMETRY AND CONSERVATION LAWS

In classical mechanics there is a well-known association between the existence of geometrical symmetries in a mechanical system and the constancy in time of certain dynamical quantities. For example, if a system is symmetric under rotations, e.g. a particle moving in a central potential, the angular momentum is conserved. If it is symmetric under translations in the x direction then p_x, the x component of linear momentum, is conserved. And so on.

These ideas find a close parallel in quantum mechanics. Consider the expectation value of some dynamical quantity α with associated operator $\hat{A}$ in a general quantum state ψ. Does the expectation value of α change with time? To find out, we compute $d\langle \hat{A} \rangle/dt$. Writing this out in full,

$$
\frac{d}{dt}\langle \psi | \hat{A} | \psi \rangle \equiv \frac{d}{dt}\int_{-\infty}^{\infty} \psi^* \hat{A}\psi \, dx
$$

$$
= \int_{-\infty}^{\infty} \frac{\partial \psi^*}{\partial t}\hat{A}\psi \, dx + \int_{-\infty}^{\infty} \psi^* \hat{A}\frac{\partial \psi}{\partial t} \, dx \qquad (6.60)
$$

for a one-dimensional system.

The Schrödinger equation (6.31) and its complex conjugate reduce the right-hand side of equation (6.60) to

$$
\frac{i}{\hbar}\int_{-\infty}^{\infty} [(\hat{H}\psi)^*(\hat{A}\psi) - \psi^* \hat{A}\hat{H}\psi] \, dx \qquad (6.61)
$$

Now interchange $(\hat{A}\psi)$ and $(\hat{H}\psi)^*$ (allowed because they are ordinary functions) in the first term. Using the fact that $\hat{H}$ is Hermitian, equation (6.36) tells us one can replace $(\hat{A}\psi)(\hat{H}\psi)^*$ by $\psi^* \hat{H}(\hat{A}\psi)$ $= \psi^* \hat{H}\hat{A}\psi$. Making these changes enables the integrand of equation (6.61) to be reduced to $\psi^*(\hat{H}\hat{A} - \hat{A}\hat{H})\psi$ or, equivalently, $\psi^*[\hat{H}, \hat{A}]\psi$. We therefore arrive at the important equation

$$
\frac{d}{dt}\langle \psi | \hat{A} | \psi \rangle = \frac{i}{\hbar}\langle \psi | [\hat{H}, \hat{A}] | \psi \rangle \qquad (6.62)
$$

This result can easily be generalized to three dimensions and to off-diagonal matrix elements $\langle \phi | \hat{A} | \psi \rangle$.

If an operator $\hat{A}$ commutes with the Hamiltonian $\hat{H}$ then the right-hand side of equation (6.62) vanishes, so $\langle \psi | \hat{A} | \psi \rangle = $ constant. Thus the expectation value of α is conserved.

When will $\hat{A}$ commute with $\hat{H}$? Recall that $\hat{H} \equiv \hat{p}^2/2m + V(r)$. Suppose $V(r)$ does not depend on some variable, say x. (That is $V(r) = V(y, z)$.) Then, because $\hat{p}_x$ commutes with $\hat{p}^2, \hat{y}$ and $\hat{z}$, it will commute with $\hat{H}$. It follows that $\langle \hat{p}_x \rangle$, i.e. the expectation value of the x component of momentum, is conserved. The fact that V, and hence $\hat{H}$, is independent of x may be expressed by saying that the system is symmetric under translations in the x direction. Thus the close relationship between symmetry and conservation laws is readily apparent and closely analogous to classical mechanics.

It will be remembered that operators which commute are called compatible. We conclude that an observable α whose operator $\hat{A}$ is compatible with the Hamiltonian has a conserved expectation value. Compatible observables possess common eigenfunctions. Therefore stationary states, i.e. eigenstates of $\hat{H}$, may be chosen to be eigenstates of such an observable α, and the states may then be labelled by both E and the eigenvalues of $\hat{A}$. A good example of this is a spherically symmetric system such as the hydrogen atom, for which angular momentum is conserved. The stationary states are denoted ψ_{nlm} (or $|nlm\rangle$) where n, the principal quantum number, labels the energy E of the levels, while l and m label the angular momentum quantum numbers associated with θ and ϕ (section 8.3).

6.6 SUMMARY

Quantum mechanics is founded on the following postulates.

1. Each physical system is described by a state function ψ which contains all the physical information about the system that can be observed and satisfies the Schrödinger equation.

2. Every observable can be represented by a Hermitian operator, the eigenvalues of which are the various possible values of the observable that would be obtained on measurement. Immediately after a measurement the state of the system is the corresponding eigenstate associated with that eigenvalue.

3. Although the choice of operators is not unique, they must comply with the commutation relations and bear the same functional relationship to each other as the corresponding classical quantities (the correspondence principle). In practice the choice of r and

$-i\hbar \mathbf{V}$ for position and momentum is conventional. Most other operators follow from these.

4. If the state function ψ is a superposition of eigenstates ψ_n corresponding to some particular observable α, i.e.

$$\psi = \sum_n c_n \psi_n$$

then a measurement of α will yield the eigenvalue λ_n belonging to the eigenstate ψ_n with probability $|c_n|^2$.

These postulates lead to an expression for the expectation value of an observable α with operator $\hat{A}$ for the state ψ,

$$\langle \psi | \hat{A} | \psi \rangle \equiv \int \psi^* \hat{A} \psi \, d\tau$$

and to the fact that compatible observables possess common eigenstates and are represented by commuting operators.

With these essential facts from the formal theory of quantum mechanics one is in a position to solve a wide variety of problems in chemistry and atomic, nuclear, molecular and solid state physics.

7

Angular momentum

One of the most important applications of the quantum mechanical formalism developed in the previous chapter is to angular momentum. This is because many microsystems are spherically symmetric, or approximately so, and angular momentum is therefore conserved. We shall also see that a careful study of quantized angular momentum leads to a completely new physical phenomenon – intrinsic spin.

7.1 OPERATORS AND COMMUTATORS

The classical definition of the angular momentum vector $\boldsymbol{L}$ is

$$\boldsymbol{L} = \boldsymbol{r} \times \boldsymbol{p} \tag{7.1}$$

and, recalling the correspondence principle, we adhere to this functional relation in quantum mechanics too. Thus, inserting the operator forms for $\hat{\boldsymbol{r}}$ and $\hat{\boldsymbol{p}}$, we obtain for the Cartesian components of $\hat{\boldsymbol{L}}$

$$\hat{L}_x = \hat{y}\hat{p}_z - \hat{z}\hat{p}_y = -i\hbar\left(y\frac{\partial}{\partial z} - z\frac{\partial}{\partial y} \right) \tag{7.2}$$

$$\hat{L}_y = \hat{z}\hat{p}_x - \hat{x}\hat{p}_z = -i\hbar\left(z\frac{\partial}{\partial x} - x\frac{\partial}{\partial z} \right) \tag{7.3}$$

$$\hat{L}_z = \hat{x}\hat{p}_y - \hat{y}\hat{p}_x = -i\hbar\left(x\frac{\partial}{\partial y} - y\frac{\partial}{\partial x} \right) \tag{7.4}$$

In spherical polar coordinates (r, θ, ϕ)

$$\hat{L}_x = i\hbar\left(\sin\phi\frac{\partial}{\partial\theta} + \cot\theta\cos\phi\frac{\partial}{\partial\phi} \right) \tag{7.5}$$

$$\hat{L}_y = i\hbar\left(-\cos\phi\frac{\partial}{\partial\theta} + \cot\theta\sin\phi\frac{\partial}{\partial\phi}\right) \tag{7.6}$$

$$\hat{L}_z = -i\hbar\frac{\partial}{\partial\phi} \tag{7.7}$$

An important quantity is $L^2 \equiv L_x^2 + L_y^2 + L_z^2$. We find

$$\hat{L}^2 = -\hbar^2\left[\frac{1}{\sin\theta}\frac{\partial}{\partial\theta}\left(\sin\theta\frac{\partial}{\partial\theta}\right) + \frac{1}{\sin^2\theta}\frac{\partial^2}{\partial\phi^2}\right] \tag{7.8}$$

It is worth noting that the term in square brackets is just the angular part of $r^2\nabla^2$ is spherical polar coordinates.

In classical mechanics one can measure all three components of L simultaneously. However, in quantum mechanics they are not compatible. We find

$$\begin{aligned}
[\hat{L}_x, \hat{L}_y] &= (\hat{y}\hat{p}_z - \hat{z}\hat{p}_y)(\hat{z}\hat{p}_x - \hat{x}\hat{p}_z) - (\hat{z}\hat{p}_x - \hat{x}\hat{p}_z)(\hat{y}\hat{p}_z - \hat{z}\hat{p}_y) \\
&= (\hat{y}\hat{p}_x - \hat{x}\hat{p}_y)(\hat{p}_z\hat{z} - \hat{z}\hat{p}_z) \\
&= -\hat{L}_z[\hat{p}_z, \hat{z}] \\
&= i\hbar\hat{L}_z
\end{aligned} \tag{7.9}$$

using the commutation relation equation (6.59) for the z components. Similarly

$$[\hat{L}_y, \hat{L}_z] = i\hbar\hat{L}_x \tag{7.10}$$

$$[\hat{L}_z, \hat{L}_x] = i\hbar\hat{L}_y \tag{7.11}$$

Recalling the relationship between simultaneous measurability and commutativity we deduce that only one of the three components of L can be precisely determined at any one time. If the system is in an eigenstate of, say, $\hat{L}_x$ it cannot also be in an eigenstate of $\hat{L}_y$ or $\hat{L}_z$.

However, proceeding as for equation (7.9), we find

$$[\hat{L}^2, \hat{L}_x] = [\hat{L}^2, \hat{L}_y] = [\hat{L}^2, \hat{L}_z] = 0 \tag{7.12}$$

so that the square of the total angular momentum may be determined simultaneously with any *one* component of L. It is conventional to choose the orientation of the axes so that the particular component which we wish to be precisely determined is L_z because this has the simplest form in spherical polar coordinates (equation (7.7)).

7.2 EIGENFUNCTIONS AND EIGENVALUES

It follows from section 6.4 that there exist functions which are simultaneously eigenfunctions of both $\hat{L}^2$ and $\hat{L}_z$ because $\hat{L}^2$ and $\hat{L}_z$

commute. To determine these functions we must solve the two eigen-value equations

$$\hat{L}^2\psi(\theta, \phi) = \lambda\psi(\theta, \phi) \qquad (7.13)$$

$$\hat{L}_z\psi(\theta, \phi) = \mu\psi(\theta, \phi) \qquad (7.14)$$

The second equation is easy. Using equation (7.7) we get

$$-i\hbar\frac{\partial\psi}{\partial\phi} = \mu\psi \qquad (7.15)$$

so

$$\psi = \Theta(\theta)e^{i\mu\phi/\hbar} \qquad (7.16)$$

where Θ is a function of θ only. The eigenvalues μ are restricted, however, by the requirement that ψ be a single-valued function of the angular coordinate ϕ:

$$\psi(\phi + 2\pi) = \psi(\phi)$$

This demands

$$\psi = \Theta e^{im\phi}, \quad m = 0, \pm 1, \pm 2, \ldots \qquad (7.17)$$

and

$$\mu = m\hbar \qquad (7.18)$$

So once again we see how a boundary condition on the wave function leads to a quantization condition on a physical quantity, in this case L_z.

The function $\Theta(\theta)$ is determined by solving equation (7.13). Substituting the form of solution (7.17) into equation (7.13), using equation (7.8) for $\hat{L}^2$, gives an equation for Θ

$$\frac{1}{\sin\theta}\frac{d}{d\theta}\left(\sin\theta\frac{d\Theta}{d\theta}\right) + \left(\frac{\lambda}{\hbar^2} - \frac{m^2}{\sin^2\theta}\right)\Theta = 0 \qquad (7.19)$$

Putting $z = \cos\theta$, equation (7.19) becomes

$$\frac{d}{dz}\left[(1 - z^2)\frac{dP}{dz}\right] + \left(\frac{\lambda}{\hbar^2} - \frac{m^2}{1 - z^2}\right)P = 0 \qquad (7.20)$$

where $P(z) = \Theta(\cos\theta)$. Equation (7.20) is well-known to mathe-maticians as the associated Legendre equation. It may readily be solved by the series method. Full details are given in Appendix B. Here we need merely note that polynomial solutions will only exist if the coefficient of each power of z, after substitution of the polynomial into equation (7.20), vanishes. Non-polynomial solutions also exist but blow up at $\theta = 0, \pi$ and must be discarded. If the leading power of P is z^n, substitution into equation (7.20) shows that the leading power of the left-hand side is z^{n+2}, with coefficient $n(n-1) - \lambda/\hbar^2$. This must vanish, so $\lambda = n(n-1)\hbar^2$. It is conventional to put $n = l+1$ and write

$$\lambda = l(l+1)\hbar^2, \quad l = 0, 1, 2, \ldots \tag{7.21}$$

Obviously $\langle \hat{L}_z^2 \rangle$ cannot exceed $\langle \hat{L}^2 \rangle$ so we must have $m^2 \leqslant l(l+1)$ which requires

$$-l \leqslant m \leqslant l$$

a condition which also emerges automatically from the detailed solution of equation (7.20). The functions P must therefore be labelled by the $\hat{L}^2$ eigenvalue parameter l, and also the $\hat{L}_z$ eigenvalue parameter m; thus, P_l^m.

Special interest attaches to the reflection properties of these eigenfunctions. In three dimensions the *parity operator* $\hat{P}$ changes $\mathbf{r} \to -\mathbf{r}$, or $\theta \to \pi - \theta$, $\phi \to \phi + \pi$. This operation leaves $\hat{L}^2$ and $\hat{L}_z$ unchanged, so

$$[\hat{P}, \hat{L}^2] = [\hat{P}, \hat{L}_z] = 0$$

and we conclude that the angular momentum eigenstates are also parity eigenstates. Inspection of equation (7.17) shows that, under reflection, $e^{im\phi} \to e^{im(\phi + \pi)} = (-1)^m e^{im\phi}$, indicating a parity $(-1)^m$; even $(+)$ for even m, odd $(-)$ for odd m. The parity of $P_l^m(\cos\theta)$ is $(-1)^{l-|m|}$.

Combining together the ϕ- and θ-dependent factors, we arrive at the simultaneous eigenfunctions of $\hat{L}^2$ and $\hat{L}_z$:

$$|l, m\rangle \equiv \psi_{lm}(\theta, \phi) = N_{lm} P_l^m(\cos\theta) e^{im\phi} \tag{7.22}$$

where the normalization factor

$$N_{lm} = (-1)^m \left[\frac{2l+1}{4\pi} \frac{(l-m)!}{(l+m)!} \right]^{1/2} \tag{7.23}$$

These eigenfunctions are not unique to quantum mechanics. They occur in a wide variety of problems in wave theory and potential theory, and go by the name of *spherical harmonics*, denoted $Y_{lm}(\theta, \phi)$. In Table 7.1 a few have been listed explicitly. Their parity is determined by the product of the parities of P_l^m and $e^{im\phi}$, which is $(-1)^m(-1)^{l-|m|} = (-1)^l$. Spherical harmonics have many interesting and useful properties, and they are discussed in more detail in Appendix B. Figure 7.1 illustrates some of the character of the functions as polar plots of their squared moduli $|Y_{lm}(\theta, \phi)|^2 =$

Table 7.1 *Some low-order spherical harmonics*

$Y_{00} = (4\pi)^{-1/2}$

$Y_{10} = (3/4\pi)^{1/2} \cos\theta$

$Y_{1\pm1} = \mp (3/8\pi)^{1/2} \sin\theta\, e^{\pm i\phi}$

$Y_{20} = (5/16\pi)^{1/2}(3\cos^2\theta - 1)$

$Y_{2\pm1} = \mp (15/8\pi)^{1/2} \cos\theta \sin\theta\, e^{\pm i\phi}$

$Y_{2\pm2} = (15/32\pi)^{1/2} \sin^2\theta\, e^{\pm 2i\phi}$

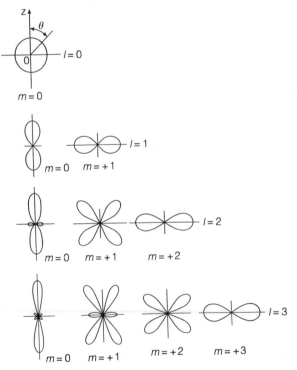

Fig. 7.1 These are polar plots of the squared moduli of Y_{lm} (tabulated in Table 7.1 for $l = 0, 1, 2, 3$). The plots, which are angular probability distributions, show an increasingly complicated pattern of lobes for higher values of m and/or l.

$|\Theta_{lm}(\theta)|^2/2\pi$ versus θ, where θ is measured from the z axis as usual. Also we note the orthonormality condition

$$\langle l, m | l', m' \rangle = \int_0^{2\pi} d\phi \int_0^{\pi} Y_{lm}^*(\theta, \phi) Y_{l'm'}(\theta, \phi) \sin\theta \, d\theta$$

$$= \delta_{ll'} \delta_{mm'}$$

We can now return to equations (7.13) and (7.14) and explicitly insert the eigenfunctions ψ and eigenvalues λ and μ. Using equations (7.18), (7.21) and (7.22) we obtain

$$\hat{L}^2 Y_{lm} = l(l+1)\hbar^2 Y_{lm} \tag{7.24}$$

$$\hat{L}_2 Y_{lm} = m\hbar Y_{lm} \tag{7.25}$$

Knowing the two values $\langle \hat{L}^2 \rangle$ and $\langle \hat{L}_z \rangle$ for a particular eigenfunction we can deduce the angle of the vector L from the z axis. Because the values of these quantities are discrete, L cannot point in any direction but must assume certain definite orientations. For a given value of l there will be $2l + 1$ allowed directions, as m ranges over the $2l + 1$ integers from $-l$ to $+l$ ($-l \leqslant m \leqslant l$). Note that m^2 is always less than $l(l+1)$, so L can never point along the z axis itself. The cases $l = 1, 2$ and 3 are shown in Fig. 7.2.

The fact that the angular momentum vector of a quantum particle must point in certain restricted directions is a bizarre quantum property. In macroscopic systems $\langle \hat{L}^2 \rangle$, and hence l, is so large that the billions of allowed angles are indistinguishable from a continuum, but in atomic systems this angle quantization leads to some important effects.

Although the polar angle is fixed for a given angular momentum

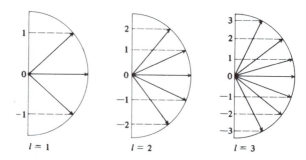

Fig. 7.2 The effective orientation of the L vector relative to the (arbitrary) z axis is quantized to $2l + 1$ discrete angles.

eigenstate, L_x and L_y are indeterminate, so the azimuthal angle about the z axis does not take on a well-defined value. We cannot say where on the x–y plane the projection of L will be. At first this seems to accord privileged status to the z direction, but we could equally well have constructed simultaneous eigenfunctions of $\hat{L}^2$ or $\hat{L}_x$ or $\hat{L}_y$, leaving $\hat{L}_z$ indeterminate. What is the case is that if we *select a direction* for the determination of one component of L (call it L_z or anything else) then by virtue of the measurements we must make in order to measure that component, we do single out a privileged direction in space. This is another example of how the quantum reality cannot be separated from the observer's choice of measurement strategy.

7.3 MATRIX FORMULATION

In section 6.2 it was shown that we may associate Hermitian matrices with Hermitian operators, and that the algebra of the operators is parallelled by the matrix algebra. We saw how the essential content of quantum mechanics is contained in the commutation relations, and that the use of differential operators and wave functions was just one of many possible representations of the commutator algebra. Another representation is to use matrices. In the case of angular momentum the matrix formulation is very instructive.

The matrices can be derived either directly from the commutator algebra or from the wave functions $\psi_{lm}\,(= Y_{lm})$ by the prescription of equation (6.41). There is, however, an important difference. While the wave functions will certainly always generate matrices that satisfy the correct angular momentum commutation relations, it need not follow that *every* matrix representation of those relations will be so generated. Let us see.

From equation (6.41) we can construct

$$\int Y_{lm}^*(\theta, \phi)\,\hat{A}\,Y_{l'm'}(\theta, \phi)\,\mathrm{d}\Omega \qquad (7.26)$$

where $\hat{A}$ is an operator, say $\hat{L}^2$, or $\hat{L}_z$, and $\mathrm{d}\Omega$ is a solid angle ($\mathrm{d}\Omega \equiv \sin\theta\,\mathrm{d}\theta\,\mathrm{d}\phi$). Now equation (7.26) carries four labels, l, l', m, m', whereas a matrix usually has only two, one to label the rows, the other the columns. Because in three dimensions we have both θ and ϕ angular degrees of freedom we really have a 'matrix of matrices', one matrix for each pair of values of l and l'. However it is easy to show that for L operators, equation (7.26) vanishes unless $l = l'$, so for non-vanishing matrices l and l' cannot be chosen independently. Each value of l leads to a matrix which has the right algebraic properties to represent a Hermitian operator. This procedure therefore generates many, indeed an infinite number of, representations of the angular momentum operators in matrix form. For a given value of l, m and m' both run over $2l + 1$ values $(-l \leqslant m, m' \leqslant l)$ so we get $(2l + 1)$-sided square matrices.

The lowest non-trivial order is for $l = 1$. Evaluating the $\hat{L}_z$ matrix for this order:

$$-i\hbar \int Y_{lm}^*(\theta, \phi) \frac{\partial}{\partial \phi} Y_{lm'}(\theta, \phi) \sin \theta \, d\theta \, d\phi, \quad m, m' = 0, \pm 1 \quad (7.27)$$

With the help of Appendix A we obtain

$$\hat{L}_z = \hbar \begin{bmatrix} 1 & 0 & 0 \\ 0 & 0 & 0 \\ 0 & 0 & -1 \end{bmatrix} \quad (7.28)$$

As expected from the discussion on pages 78–9, the matrix is diagonal (because the Y_{lm} are eigenfunctions of $\hat{L}_z$) and the diagonal components are the eigenvalues (equation (7.18). The matrix of $\hat{L}^2$ will also be diagonal. Using equation (7.8) in place of equation (7.7) in integral (7.27) we get

$$\hat{L}^2 = 2\hbar^2 \begin{bmatrix} 1 & 0 & 0 \\ 0 & 1 & 0 \\ 0 & 0 & 1 \end{bmatrix} \quad (7.29)$$

where once again the diagonal matrix displays the eigenvalues of $\hat{L}^2$, i.e. $l(l + 1)\hbar^2 = 2\hbar^2$ in this case.

The matrices for $\hat{L}_x$ and $\hat{L}_y$ are obtained similarly:

$$\hat{L}_x = \frac{\hbar}{\sqrt{2}} \begin{bmatrix} 0 & 1 & 0 \\ 1 & 0 & 1 \\ 0 & 1 & 0 \end{bmatrix}, \quad \hat{L}_y = \frac{\hbar}{\sqrt{2}} \begin{bmatrix} 0 & -i & 0 \\ i & 0 & -i \\ 0 & i & 0 \end{bmatrix} \quad (7.30)$$

and are not diagonal because the Y_{lm} are not eigenfunctions of $\hat{L}_x$ and $\hat{L}_y$. Note, however, that all these matrices are manifestly Hermitian.

It is now a straightforward matter to verify that the above matrices correctly satisfy the various commutation relations given in the previous section. For example

$$[\hat{L}_x, \hat{L}_y] = \frac{\hbar}{2}\begin{bmatrix} 0 & 1 & 0 \\ 1 & 0 & 1 \\ 0 & 1 & 0 \end{bmatrix}\begin{bmatrix} 0 & -i & 0 \\ i & 0 & -i \\ 0 & i & 0 \end{bmatrix}$$

$$-\frac{\hbar^2}{2}\begin{bmatrix} 0 & -i & 0 \\ i & 0 & -i \\ 0 & i & 0 \end{bmatrix}\begin{bmatrix} 0 & 1 & 0 \\ 1 & 0 & 1 \\ 0 & 1 & 0 \end{bmatrix}$$

$$= i\hbar^2 \begin{bmatrix} 1 & 0 & 0 \\ 0 & 0 & 0 \\ 0 & 0 & -1 \end{bmatrix}$$

$$= i\hbar\hat{L}_z$$

Being both diagonal, the $\hat{L}^2$ and $\hat{L}_z$ matrices clearly commute.

One can now go on to construct the 5×5 matrices corresponding to $l = 2$, and so on. At each order the $\hat{L}_z$ and $\hat{L}^2$ matrices will be diagonal with components given by the eigenvalues, equations (7.18) and (7.21) respectively. The others will be more complicated. They will all satisfy the same commutator algebra as the $l = 1$ representation above.

7.4 INTRINSIC SPIN

It will be noticed that all the matrices discussed above have an odd number of rows and columns, and the thought naturally occurs as to whether there exist even-component matrices (with 2, 4, 6,... rows and columns) which also satisfy the angular momentum commutation, but have been missed by the prescription given by (7.26). In particular, the lowest-order representation so far is 3×3, but a 2×2 matrix is still a non-trivial operator and there seems a possibility that a 2×2 representation of the angular momentum operators exists. In fact it does:

$$\hat{L}_x = \tfrac{1}{2}\hbar\begin{bmatrix} 0 & 1 \\ 1 & 0 \end{bmatrix}, \quad \hat{L}_y = \tfrac{1}{2}\hbar\begin{bmatrix} 0 & -i \\ i & 0 \end{bmatrix}, \quad \hat{L}_z = \tfrac{1}{2}\hbar\begin{bmatrix} 1 & 0 \\ 0 & -1 \end{bmatrix} \qquad (7.31)$$

$$\hat{L}^2 = \tfrac{3}{4}\hbar^2\begin{bmatrix} 1 & 0 \\ 0 & 1 \end{bmatrix} \qquad (7.32)$$

corresponding to $l = \tfrac{1}{2}$, $m = \pm\tfrac{1}{2}$. This result is readily achieved by taking arbitrary Hermitian 2×2 matrices and applying the commutation relations.

A more instructive method, however, is to deduce the eigenvalue spectrum for l and m directly from the commutation relations. This can be accomplished using a method analogous to the step-up and step-down operators used to deduce the energy eigenvalues of the simple harmonic oscillator in section 4.1. The analogous operators here are

$$\hat{L}_{\pm} = \hat{L}_x \pm i \hat{L}_y \tag{7.33}$$

($\hat{L}_{\pm}$ are not Hermitian). Using equations (7.9) to (7.12) we get

$$[\hat{L}_z, \hat{L}_{\pm}] = \pm \hbar \hat{L}_{\pm} \tag{7.34}$$

$$[\hat{L}^2, \hat{L}_{\pm}] = 0 \tag{7.35}$$

$$\hat{L}_{\pm}\hat{L}_{\mp} = \hat{L}^2 - \hat{L}_z^2 \pm \hbar \hat{L}_z \tag{7.36}$$

Returning, then, to equations (7.13) and (7.14) without prejudice concerning the values of λ and μ we have from equation (7.35)

$$\hat{L}_{\pm}\hat{L}^2\psi = \hat{L}^2(\hat{L}_{\pm}\psi) = \lambda(\hat{L}_{\pm}\psi) \tag{7.37}$$

so $\hat{L}_{\pm}\psi$ is an eigenvector of $\hat{L}^2$. From equation (7.34)

$$\hat{L}_+\hat{L}_z\psi = \mu(\hat{L}_+\psi) = \hat{L}_z(\hat{L}_+\psi) - \hbar(\hat{L}_+\psi)$$

so

$$\hat{L}_z(\hat{L}_+\psi) = (\mu + \hbar)(\hat{L}_+\psi) \tag{7.38}$$

and we see that $\hat{L}_+$ takes an eigenvector of $\hat{L}_z$ and steps it up to another eigenvector, $\hat{L}_+\psi$, with eigenvalue $\hbar$ greater. Similarly $\hat{L}_-$ steps down by one unit of $\hbar$.

As in the oscillator case, one cannot go on stepping for ever. Here the value of m is hemmed in both ways by the restriction $m^2 \leqslant l^2$, so sooner or later there has to be an eigenvector, call it ψ_t (t for top), where

$$\hat{L}_+\psi_t = 0 \tag{7.39}$$

and the ladder ends. Likewise, there is a ψ_b (b for bottom)

$$\hat{L}_-\psi_b = 0 \tag{7.40}$$

This restriction gives us the eigenvalues right away because from equations (7.36) and (7.39), $\hat{L}_x\hat{L}_+\psi_t = (\hat{L}^2 - \hat{L}_z^2 - \hbar\hat{L}_z)\psi_t = 0$, so

$(\lambda - \mu_t^2 - \hbar\mu_t)\psi_t = 0$, which implies

$$\lambda = \mu_t(\mu_t + \hbar) \tag{7.41}$$

for the corresponding top eigenvalue μ_t.

Similarly from equation (7.40)

$$\lambda = \mu_b(\mu_b - \hbar) \tag{7.42}$$

and equations (7.41) and (7.42) together require $\mu_t = -\mu_b$. We know that the steps between μ_b and μ_t are whole units of $\hbar$ so

$$\mu_t - \mu_b = m\hbar \qquad m = 0, 1, 2, 3, \ldots$$

We conclude

$$\mu_t = -\mu_b = (m/2)\hbar \tag{7.43}$$

and from equation (7.41)

$$\lambda = l(l+1)\hbar^2 \qquad l = 0, \tfrac{1}{2}, 1, \tfrac{3}{2}, 2, \ldots \tag{7.44}$$

Thus the straightforward use of the commutation relations, equations [7.9] to [7.12], has led us to an eigenvalue spectrum in which angular momentum is quantized in *half-integral* units of $\hbar$. The lowest non-trivial case is $l = \tfrac{1}{2}$, $m = \pm\tfrac{1}{2}$, which corresponds to the matrices, equations (7.31) and (7.32). From the form of $\hat{L}_\pm$ one may reconstruct $\hat{L}_x$ and $\hat{L}_y$. Clearly, in addition to the 2×2 representation, there exist representations for all even values of $2l + 1$. All these cases were missed by the procedure of starting with the eigenfunctions Y_{lm}. Why?

The quantization of μ and λ was achieved in the previous section through the boundary conditions imposed on the wave function to make ψ single valued in ϕ and finite at $\theta = 0, \pi$. This ought to give us all the eigenvalues for any sort of angular momentum derived from the definition in equation (7.1). The half-integral eigenvalues, then, must correspond to a type of angular momentum not included in this definition. Inspection of equation (7.1) shows that L is built out of the position of the particle r and its motion through p. Could there exist a type of angular momentum that is not dependent on these variables? If so it must be somehow *internal* to the particle, a sort of interior degree of freedom. Such an angular momentum would not be involved in the imposition of spatial boundary conditions on the wave function ψ because it is 'inside' the particle and not dependent on the particle's location or motion. That is why we missed it in the wave function quantization.

In some respects the abstract idea of an internal angular momentum is reminiscent of the classical concept of a spinning body. The Earth, for example, has orbital angular momentum – the $r \times p$ variety – (motion around the sun) but also rotates on its axis and so has some extra spin angular momentum that is independent of its position or orbital motion. For this reason the half-integral angular momentum is known as *intrinsic spin*, or just 'spin'. The analogy is only partly successful, however, because the spin of the Earth is still an $r \times p$ angular momentum, but r and p now referring to bits of Earth going around its axis rather than to the Earth's centre of gravity going round the sun. Intrinsic spin differs in the fundamental respect of its half-integral values. This leads to weird physical effects that simply cannot be realized by any type of rotating classical body.

There still remains the question of whether intrinsic spin, whatever its mathematical plausibility, is actually realized in nature. Historically its existence was originally suspected from the structure of certain spectral lines and the idea was introduced for electrons on an *ad hoc* basis by Goudsmidt and Uhlenbeck in 1929. Shortly thereafter Dirac, in attempting to reconcile quantum mechanics with the special theory of relativity, produced a relativistic wave equation to replace Schrödinger's and showed that spin $l = \frac{1}{2}$ was automatically present in his equation. This encouraged the view that spin could be regarded as an essentially relativistic effect.

Today we know that electrons, protons, neutrons, neutrinos and many other subatomic particles possess spin one-half, and yet other particles, e.g. the so-called Ω^-, have spin $l = \frac{3}{2}$. Not all particles fall into this category though. Pions have no spin, while photons and some mesons have spin $l = 1$. These do not obey Dirac's equation. The physical properties of these two distinct classes of particles are very different. In particular the half-integral, i.e. intrinsic, spin particles obey the Pauli exclusion principle (see section 4.6).

7.5 SPIN EIGENVECTORS

Before finding the eigenvectors belonging to the spin angular momentum matrices a couple of notational issues must be stated. First, the matrices in equations (7.31) and (7.32) are customarily denoted σ and referred to as the Pauli spin matrices

$$\hat{\sigma}_x = \begin{bmatrix} 0 & 1 \\ 1 & 0 \end{bmatrix}, \quad \hat{\sigma}_y = \begin{bmatrix} 0 & -i \\ i & 0 \end{bmatrix}, \quad \hat{\sigma}_z = \begin{bmatrix} 1 & 0 \\ 0 & -1 \end{bmatrix} \quad (7.45)$$

Second, the angular momentum operators $\hat{L}_x$, etc., constructed from equation (7.45), are frequently denoted by S rather than L to remind us of the distinction between intrinsic spin and 'ordinary' angular momentum L.

The common eigenvectors of $\hat{S}_z$ and $\hat{S}^2$ are found by solving the eigenvalue equations

$$\left.\begin{aligned}\hat{S}_z\begin{bmatrix}a\\b\end{bmatrix}&\equiv\tfrac{1}{2}\hbar\begin{bmatrix}1&0\\0&-1\end{bmatrix}\begin{bmatrix}a\\b\end{bmatrix}=\mu\begin{bmatrix}a\\b\end{bmatrix}\\[2mm]\hat{S}^2\begin{bmatrix}a\\b\end{bmatrix}&\equiv\tfrac{3}{4}\hbar^2\begin{bmatrix}1&0\\0&1\end{bmatrix}\begin{bmatrix}a\\b\end{bmatrix}=\lambda\begin{bmatrix}a\\b\end{bmatrix}\end{aligned}\right\}\tag{7.46}$$

for eigenvalues $\lambda = 3\hbar^2/4$ and $\mu = \pm\,\hbar/2$. This gives

$$\left.\begin{aligned}\begin{bmatrix}a\\b\end{bmatrix}&=\begin{bmatrix}1\\0\end{bmatrix},\quad\mu=+\tfrac{1}{2}\hbar\\[2mm]\begin{bmatrix}a\\b\end{bmatrix}&=\begin{bmatrix}0\\1\end{bmatrix},\quad\mu=-\tfrac{1}{2}\hbar\end{aligned}\right\}\tag{7.47}$$

The general object $\begin{bmatrix}a\\b\end{bmatrix}$ is known as a *spinor*.

The spin vector S can point at only one of two allowed angles to the z axis (Fig. 7.3). Although S cannot actually point along the z axis, becase $\langle\hat{S}_z^2\rangle < \langle\hat{S}^2\rangle$, these two allowed directions are still referred to as 'spin up' and 'spin down'. It is then convenient to write $|\uparrow\rangle$ and $|\downarrow\rangle$ for the eigenvectors belonging to $+\hbar/2$ and $-\hbar/2$ respectively.

The eigenvalues of $\hat{S}_x$ and $\hat{S}_y$ are also $\pm\hbar/2$. The eigenvectors are

$$\hat{S}_x:\frac{1}{\sqrt{2}}\begin{bmatrix}1\\1\end{bmatrix},\quad+\hbar/2;\quad\frac{1}{\sqrt{2}}\begin{bmatrix}-1\\1\end{bmatrix},\quad-\hbar/2\tag{7.48}$$

$$\hat{S}_y:\frac{1}{\sqrt{2}}\begin{bmatrix}1\\i\end{bmatrix},\quad+\hbar/2;\quad\frac{1}{\sqrt{2}}\begin{bmatrix}1\\-i\end{bmatrix},\quad-\hbar/2\tag{7.49}$$

The scalar product $\langle\,|\,\rangle$ is here an ordinary scalar product of vectors, except the entries in the 'bra' $\langle\,|$ are complex conjugated. Thus, for example, normalization requires

$$\langle\psi|\psi\rangle=[a^*\;\;b^*]\begin{bmatrix}a\\b\end{bmatrix}=|a|^2+|b|^2=1\tag{7.50}$$

Using this definition it is easy to verify the orthonormality of all the

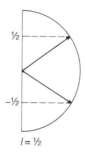

Fig. 7.3 The spin vector S is quantized to two discrete angles, referred to as 'spin up' and 'spin down'.

eigenvectors in equations (7.47) and (7.49). Note that for half-integer spin we are obliged to use matrices and vectors; there is no equivalent to the alternative eigenfunctions Y_{lm} that we can use in the case of integer l.

7.6 ILLUSTRATING THE RULES OF QUANTUM MECHANICS

Because a system with intrinsic spin, in which the motion of the system through space is ignored, has only two angular momentum eigenstates it is a particularly simple system with which to illustrate some of the basic principles of quantum mechanics presented in the previous chapter.

A general quantum mechanical state $|\psi\rangle$ will be a linear super-position of the eigenstates $|\uparrow\rangle$ and $|\downarrow\rangle$:

$$|\psi\rangle = c_1|\uparrow\rangle + c_2|\downarrow\rangle \quad \text{or} \quad c_1\begin{bmatrix} 1 \\ 0 \end{bmatrix} + c_2\begin{bmatrix} 0 \\ 1 \end{bmatrix} \quad (7.51)$$

where c_1 and c_2 are complex numbers. From postulate 4 in section 6.6 we recall that $|c_1|^2$ is the probability that, on measurement of $\hat{S}_z$, the system will be found to have spin up, i.e. eigenvector $+\frac{1}{2}\hbar$, while $|c_2|^2$ is the probability for spin down $(-\frac{1}{2}\hbar)$. Clearly $|c_1|^2 + |c_2|^2 = 1$. It follows that

$$\langle \hat{S}_z \rangle \equiv \langle \psi | \hat{S}_z | \psi \rangle = \frac{1}{2}\hbar [c_1^* \ c_2^*]\begin{bmatrix} 1 & 0 \\ 0 & -1 \end{bmatrix}\begin{bmatrix} c_1 \\ c_2 \end{bmatrix}$$

$$= \frac{1}{2}\hbar(|c_1|^2 - |c_2|^2) \quad (7.52)$$

which manifestly has the form

probability of spin up × $(\hbar/2)$ + probability of spin down × $(-\hbar/2)$

Also

$$\langle \hat{S}^2 \rangle = \tfrac{3}{4}\hbar^2 [c_1^* \; c_2^*] \begin{bmatrix} 1 & 0 \\ 0 & 1 \end{bmatrix} \begin{bmatrix} c_1 \\ c_2 \end{bmatrix}$$

$$= \tfrac{3}{4}\hbar^2 (|c_1|^2 + |c_2|^2) = \tfrac{3}{4}\hbar^2 \qquad (7.53)$$

As already noted, the z direction is determined by the axis along which one chooses to measure the component of spin. It is important to realize that *whatever axis is chosen*, the only possible values that can be obtained for the component of spin along it are $\pm \tfrac{1}{2}\hbar$. This remains true even if, prior to the measurement, one can say that the spin vector certainly points along a *different* axis.

To illustrate this stange feature let us suppose that the system is initially in a spin-up eigenstate $|\uparrow\rangle$ referred to the z axis of some particular coordinate system. Now we choose to measure the component of S along some other axis, z', inclined at angle θ to the original z axis. Classically we may relate S_z and $S_{z'}$ by $S_{z'} = S_x \sin\theta + S_z \cos\theta$, and in what follows we take this to be true for the $\hat{S}$ operators too.

Using equations (7.31) we get an eigenvector equation for $S_{z'}$:

$$\hat{S}_{z'} \begin{bmatrix} a \\ b \end{bmatrix} \equiv \tfrac{1}{2}\hbar \begin{bmatrix} \cos\theta & \sin\theta \\ \sin\theta & -\cos\theta \end{bmatrix} \begin{bmatrix} a \\ b \end{bmatrix} = \mu \begin{bmatrix} a \\ b \end{bmatrix} \qquad (7.54)$$

This is readily solved to give $\mu = \pm \hbar/2$ again. Whatever direction we measure the component of spin along we always get the answer $+\hbar/2$ or $-\hbar/2$. The eigenvectors of $\hat{S}_{z'}$ are

$$|\uparrow'\rangle \equiv \begin{bmatrix} \cos(\theta/2) \\ \sin(\theta/2) \end{bmatrix}, \qquad \mu = +\hbar/2 \qquad (7.55)$$

$$|\downarrow'\rangle \equiv \begin{bmatrix} -\sin(\theta/2) \\ \cos(\theta/2) \end{bmatrix}, \qquad \mu = -\hbar/2 \qquad (7.56)$$

When $\theta = 0$, z' coincides with z, and equations (7.55) and (7.56) reduce to equations (7.47). When $\theta = \pi/2$, z' coincides with x. With this choice of z' axis equations (7.55) and (7.56) reduce to equations (7.48).

If the spin-up state $|\uparrow\rangle$, referred to the original z axis, is expanded in terms of the $\hat{S}_{z'}$ eigenstates, we get

$$|\uparrow\rangle = \cos(\theta/2)|\uparrow'\rangle - \sin(\theta/2)|\downarrow'\rangle \qquad (7.57)$$

The probability that, if the spin initially points up the z axis, a measurement of the spin component along z' will yield $+\hbar/2$ (spin 'up' the z' axis) is $\cos^2(\theta/2)$, while spin down has a probability $\sin^2(\theta/2)$. When $\theta = \pi/2$, the z' axis is orthogonal to the z axis and there is equal likelihood of finding the spin pointing either up or down the z' axis. Notice that so long as $\theta \neq 0$ there is always a chance that a measurement will cause the spin to flip from pointing up the z axis to down the z' axis. Using the matrix in equation (7.54) with the eigenvector (equation (7.57)), we find for the expectation value of $\hat{S}_{z'}$

$$\langle \uparrow | \hat{S}_{z'} | \uparrow \rangle = \tfrac{1}{2}\hbar \cos\theta$$
$$= (+\tfrac{1}{2}\hbar)\cos^2(\theta/2) + (-\tfrac{1}{2}\hbar)\sin^2(\theta/2)$$

A curious situation occurs if we put $\theta = 2\pi$, corresponding to the z' axis being rotated through $360°$ back to coincide with the z axis once again. Because of the $\theta/2$ dependence of the above expressions, we find that the eigenvectors, equations (7.55) and (7.56), are not the same as those for $\theta = 0$. In fact they are the *negatives* of $|\uparrow\rangle$ and $|\downarrow\rangle$ and so differ by a phase factor. In many cases the phase of the eigenvector is irrelevant, e.g. $\langle \hat{S}_{z'} \rangle$ is the same whether $\theta = 0$ or 2π. Nevertheless there are some experiments where phase differences might actually be measured, such as in an interference experiment. Evidently it is necessary to rotate a particle with intrinsic spin twice, i.e. through 4π, before its original physical state is restored. This weird double-valued aspect of intrinsic spin sets it apart from ordinary angular momentum (and intuition). It cautions us not to attach too literal a meaning to the word spin. The classical image of a body rotating about an axis is totally inadequate to describe the peculiar geometrical properties of intrinsic spin. The nature of the spin of a particle such as an electron has no direct counterpart in the macroscopic world of our experience.

7.7 MAGNETIC MOMENT

As in classical mechanics a charged quantum system with angular momentum possesses a magnetic moment $\boldsymbol{\mu}$. Classically

$$\boldsymbol{\mu} = -(e/2m_e)\boldsymbol{L} \tag{7.58}$$

for a charge $-e$ (we are thinking of an electron) and mass m_e. If an external magnetic field is applied there is an interaction energy of $-\boldsymbol{\mu} \cdot \boldsymbol{B}$. We assume these expressions to hold in quantum mechanics too. Choosing $\boldsymbol{B}$ to lie along the z axis, the magnetic interaction supplies an addition Hamiltonian

$$\hat{H}_{\text{mag}} = (eB/2m_e)\hat{L}_z \tag{7.59}$$

If the rest of the Hamiltonian, $\hat{H}_0$, commutes with $\hat{L}_z$ (this is the case for spherically symmetrical systems) then the stationary states of the system will be eigenstates of $\hat{L}_z$, which we denote $|m\rangle$, with eigenvalues $m\hbar$. Suppose the energy is E_0 when B is switched off:

$$\hat{H}_0|m\rangle = E_0|m\rangle \tag{7.60}$$

Then in the presence of B, equations (7.59) and (7.60) yield

$$(\hat{H}_0 + \hat{H}_{\text{mag}})|m\rangle = \left(E_0 + \frac{eBm\hbar}{2m_e} \right)|m\rangle \tag{7.61}$$

from which one sees that the energy levels are split by an energy shift

$$\Delta E = \mu_B B m \qquad m = 0, \pm 1, \pm 2, \ldots \tag{7.62}$$

where $\mu_B \equiv e\hbar/2m_e$ is known as the Bohr magneton. Thus the magnetic field splits the degenerate E_0 energy level into a multiplet of $2l + 1$ levels.

The electron also possesses an intrinsic magnetic moment due to its spin S. Dirac's relativistic equation predicts

$$\boldsymbol{\mu}_S = -(e/m_e)\boldsymbol{S} \tag{7.63}$$

which should be compared with the orbital magnetic moment given by equation (7.58). The double-valued character of spin (see above) reappears here by introducing a classically inexplicable factor of two in the magnetic moment to angular momentum ratio. An external magnetic field will split spin states into doublets with separation $2\mu_B B$.

7.8 SPIN WAVE FUNCTIONS

In a general problem we have to contend with both the spin of a particle and its motion through space. A proper treatment requires Dirac's relativistic theory. Nevertheless it is possible to incorporate some of the effects of spin into the Schrödinger equation by simply allowing the wave function $\psi(\boldsymbol{r}, t)$ to become the two-component object

$$\begin{bmatrix} \psi_1(\boldsymbol{r}, t) \\ \psi_2(\boldsymbol{r}, t) \end{bmatrix} \tag{7.64}$$

The Schrödinger equation then reads

$$\hat{H} \begin{bmatrix} \psi_1(r,t) \\ \psi_2(r,t) \end{bmatrix} = i\hbar \frac{\partial}{\partial t} \begin{bmatrix} \psi_1(r,t) \\ \psi_2(r,t) \end{bmatrix} \tag{7.65}$$

where $\hat{H}$ is now the *total* Hamiltonian, generally consisting of both space-dependent and spin-dependent pieces.

Normalization requires

$$\int (\psi_1 \psi_1^* + \psi_2 \psi_2^*)\, d\tau = 1 \tag{7.66}$$

and we interpret $|\psi_1|^2$ and $|\psi_2|^2$ as the position probability densities for a spin-up and spin-down particle respectively.

If $\hat{H}$ does not contain any explicit spin-dependent part then both ψ_1 and ψ_2 satisfy the ordinary Schrödinger equation. So long as we are not interested in the spin direction of the particles concerned we may continue to use all the previous results, ignoring the two-component nature of the wave function and working with a single ψ.

Conversely, if we can ignore the spatial motion then equation (7.65) reduces to

$$\hat{H}_S \begin{bmatrix} a(t) \\ b(t) \end{bmatrix} = i\hbar \frac{\partial}{\partial t} \begin{bmatrix} a(t) \\ b(t) \end{bmatrix} \tag{7.67}$$

where $\hat{H}_S$ is the spin-dependent part of the Hamiltonian and the spinor $\begin{bmatrix} a \\ b \end{bmatrix}$ is now explicitly time dependent. (We were able to ignore this time dependence in section 7.6 because the states $|\uparrow\rangle$ and $|\downarrow\rangle$, being degenerate, had identical time dependence.)

In many problems of interest the total angular momentum of a system is composed of several components. For example an electron may possess both spin and orbital angular momentum, or a many-electron atom will have spin contributions from several particles. In classical mechanics angular momentum is added vectorially but in the quantum case the issue is more subtle. The procedure will be illustrated by the example of the addition of two spins.

The total spin operator for two identical particles will be $\hat{S}_1 + \hat{S}_2$, where $\hat{S}_1$ acts only on the spin state associated with particle 1, while $\hat{S}_2$ acts on the spin state of particle 2. There will be four possible spin states of the two particles, according to whether each spin is directed either up or down. We can form the following combinations:

$$|\uparrow_1\rangle|\uparrow_2\rangle, \quad |\uparrow_1\rangle|\downarrow_2\rangle, \quad |\downarrow_1\rangle|\uparrow_2\rangle, \quad |\downarrow_1\rangle|\downarrow_2\rangle \tag{7.68}$$

As usual, we wish to find eigenvectors of $\hat{S}^2$ and $\hat{S}_z$, where S refers to the total angular momentum. Noting that $\hat{S}_1$ and $\hat{S}_2$ commute (because they refer to different particles) we have

$$\hat{S}^2 = \hat{S}_1^2 + \hat{S}_2^2 + 2\hat{S}_1 \cdot \hat{S}_2 \qquad (7.69)$$

We may use the explicit forms, equations (7.31) and (7.32), for the S operators in equation (7.69), then act on each of the states listed in [7.68] in turn with S^2. The result is

$$\hat{S}^2|\uparrow_1\rangle|\uparrow_2\rangle = 2\hbar^2|\uparrow_1\rangle|\uparrow_2\rangle \qquad (7.70)$$

$$\hat{S}^2|\uparrow_1\rangle|\downarrow_2\rangle = \hbar^2(|\uparrow_1\rangle|\downarrow_2\rangle + |\downarrow_1\rangle|\uparrow_2\rangle) \qquad (7.71)$$

$$\hat{S}^2|\downarrow_1\rangle|\uparrow_2\rangle = \hbar^2(|\uparrow_1\rangle|\downarrow_2\rangle + |\downarrow_1\rangle|\uparrow_2\rangle) \qquad (7.72)$$

$$\hat{S}^2|\downarrow_1\rangle|\downarrow_2\rangle = 2\hbar^2|\downarrow_1\rangle|\downarrow_2\rangle \qquad (7.73)$$

Thus in the case of equations (7.70) and (7.73) the states $|\uparrow_1\rangle|\uparrow_2\rangle$ and $|\downarrow_1\rangle|\downarrow_2\rangle$ respectively are eigenstates of $\hat{S}^2$ with eigenvalue $2\hbar^2$. However, equations (7.71) and (7.72) show that neither $|\uparrow_1\rangle|\downarrow_2\rangle$ nor $|\downarrow_1\rangle|\uparrow_2\rangle$ is an eigenstate of $\hat{S}^2$. Nevertheless it is easy to find linear combinations which are:

$$\frac{1}{\sqrt{2}}(|\uparrow_1\rangle|\downarrow_2\rangle + |\downarrow_1\rangle|\uparrow_2\rangle) \qquad (7.74)$$

is an eigenstate of $\hat{S}^2$ with eigenvalue $2\hbar^2$, while

$$\frac{1}{\sqrt{2}}(|\uparrow_1\rangle|\downarrow_2\rangle - |\downarrow_1\rangle|\uparrow_2\rangle) \qquad (7.75)$$

is an eigenstate of $\hat{S}^2$ with eigenvalue 0.

The states in equations (7.70), (7.73), (7.74) and (7.75) are all eigenstates of $\hat{S}_z$ also, as may be found by applying this operator directly to them. Their eigenvalues are $\hbar$, $-\hbar$, 0 and 0 respectively.

These results are readily interpreted physically. The triplet of states

$$|\uparrow_1\rangle|\uparrow_2\rangle \qquad\qquad \langle\hat{S}_z\rangle = +\hbar$$

$$\frac{1}{\sqrt{2}}(|\uparrow_1\rangle|\downarrow_2\rangle + |\uparrow_2\rangle|\downarrow_1\rangle) \qquad \langle\hat{S}_z\rangle = 0$$

$$|\downarrow_1\rangle|\downarrow_2\rangle \qquad\qquad \langle\hat{S}_z\rangle = -\hbar$$

corresponds to the two spins being aligned parallel to each other, making a total angular momentum of 1. This resultant vector then takes the usual three possible m values of $+1$, 0 and -1. The remaining singlet state

$$\frac{1}{\sqrt{2}}(|\uparrow_1\rangle|\downarrow_2\rangle - |\uparrow_2\rangle|\downarrow_1\rangle) \qquad \langle \hat{S}_z \rangle = 0$$

corresponds to the spins being aligned anti-parallel to each other, giving a net angular momentum of zero.

In general, two angular momenta $\hat{\boldsymbol{L}}_1$ and $\hat{\boldsymbol{L}}_2$ can be combined quantum mechanically to yield a total $\hat{\boldsymbol{L}}$ whose l values may take any value from $|l_1 - l_2|$ to $l_1 + l_2$ in integer steps. The corresponding $\hat{L}_z$ has $2l + 1$ eigenvalues. The total number of states is therefore

$$\sum_{l=|l_1-l_2|}^{l_1+l_2} (2l+1) = (2l_1+1)(2l_2+1)$$

We shall devote more attention to the question of how to combine angular momenta in Chapter 9. It is necessary to be able to do this in order to understand the inner workings of atoms and molecules and their response to applied fields. Before that, however, we have to deal with the behaviour of particles in a central potential, the subject matter of the next chapter.

8

Particle in a central potential

One of the most important problems in quantum mechanics concerns the motion of a particle in a potential of the form $V(\mathbf{r}) = V(r)$, where $r \equiv |\mathbf{r}|$. This system has the simplifying feature of spherical symmetry, which allows the angular dependence of the motion to be reduced to purely geometrical considerations and hence solved once and for all, while the physics enters only through the radial motion. Treating this problem in detail also provides a good opportunity to illustrate the quantum mechanical formalism.

8.1 SEPARATION OF RADIAL AND ANGULAR VARIABLES

Consider a single particle (spin neglected) of mass m moving in an attractive potential $V(r)$. As V is time independent, stationary states are expected. If the potential is produced by a central body of mass M, e.g. the Coulomb field produced by an atomic nucleus, then the relative motion must be described, as in classical mechanics, by the *reduced mass* $\mu = mM/(m + M)$. For $M \gg m$, $\mu \approx m$.

In spherical polar coordinates equation (2.2) becomes

$$\hat{H}u \equiv \left\{ -\frac{\hbar^2}{2\mu} \left[\frac{1}{r^2} \frac{\partial}{\partial r} \left(r^2 \frac{\partial}{\partial r} \right) + \frac{1}{r^2 \sin\theta} \frac{\partial}{\partial \theta} \left(\sin\theta \frac{\partial}{\partial \theta} \right) \right. \right.$$

$$\left. \left. + \frac{1}{r^2 \sin^2\theta} \frac{\partial^2}{\partial\phi^2} \right] + V(r) \right\} u(r, \theta, \phi) = Eu(r, \theta, \phi) \quad (8.1)$$

which looks formidable but soon simplifies. First we notice that the angular part is almost $\hat{L}^2$, the angular momentum operator (equation

(7.8)). So

$$\left[-\frac{\hbar^2}{2\mu} \frac{1}{r^2} \frac{\partial}{\partial r} \left(r^2 \frac{\partial}{\partial r} \right) + \frac{\hat{L}^2}{2\mu r^2} + V(r) \right] u = Eu \qquad (8.2)$$

The form of equation (8.2) immediately implies that

$$[\hat{H}, \hat{L}^2] = [\hat{H}, \hat{L}_z] = 0 \qquad (8.3)$$

as $\hat{L}^2$ and $\hat{L}_z$ do not depend on r or $\partial/\partial r$, and both commute with $\hat{L}^2/2\mu r^2$. Recalling section 6.5, we conclude that $\hat{L}^2$ and $\hat{L}_z$ are conserved quantities and that there exist stationary states (energy eigenstates) which will also be angular momentum eigenstates. We have already found the angular momentum eigenfunctions in section 7.2 – they are the spherical harmonics $Y_{lm}(\theta, \phi)$. So we know that u can be factorized as follows

$$u(r, \theta, \phi) = R(r) Y_{lm}(\theta, \phi) \qquad (8.4)$$

Substituting equation (8.4) into equation (8.2) yields the radial wave equation for R. The $\hat{L}^2$ operator in equation (8.2) acts on Y_{lm}, so we may use equation (7.24) to obtain

$$\left[-\frac{\hbar^2}{2\mu} \frac{1}{r^2} \frac{d}{dr} \left(r^2 \frac{d}{dr} \right) + \frac{\hbar^2 l(l+1)}{2\mu r^2} + V(r) \right] R(r) = ER(r) \qquad (8.5)$$

where $\partial/\partial r$ has been replaced by d/dr as R is a function of r only. The l-dependent term may be written $\langle \hat{L}^2 \rangle / 2\mu r^2$ and will be recognized as the quantum counterpart of the classical 'centrifugal' potential barrier $L^2/2\mu r^2$ encountered, for example, in the Kepler problem of planetary orbits. Evidently the solutions $R(r)$ will depend on l but not on the quantum number m which is absent from equation (8.5). This was expected. The energy levels ought not to depend on $\langle \hat{L}_z \rangle = m\hbar$ as the orientation of the z axis is irrelevant in a spherically symmetrical system. Thus each energy level is $(2l + 1)$-degenerate, this being the multiplicity of m values for each l.

The degeneracy explains an otherwise puzzling feature of the spherical harmonics. Although the system is spherically symmetric, $Y_{lm}(\theta, \phi)$ is not (except in the case of Y_{00}). The wave functions, and hence the position expectation value of the particle, seem to break the spherical symmetry. This is not in fact the case. If we only know the energy of the state, we must average over all degenerate angular momentum states in taking expectation values as we have no information about which m state the particle is in. It then follows from the properties of spherical harmonics that this average is independent of θ and ϕ. Thus after many measurements one would find the particle's orientation distributed at random, and spherical symmetry (in this statistical sense) would be recovered.

8.2 RADIAL SQUARE WELL

A simple example is provided by the three-dimensional version of the infinite square well

$$V(r) = 0 \qquad r < a$$
$$\quad\,\, = \infty \qquad r > a \tag{8.6}$$

Transforming to $\rho = \alpha r$, where $\alpha = (2\mu E/\hbar^2)^{1/2}$, and $R = \chi/\rho$, equation (8.5) reduces to

$$\frac{d^2\chi}{d\rho^2} + \left[1 - \frac{l(l+1)}{\rho^2} \right] \chi = 0 \tag{8.7}$$

If we have a solution χ_l of equation (8.7) for some value of l, say $l = k$, then $-\chi'_k + (k+1)\chi_k/\rho$ is a solution of equation (8.7) for $l = k + 1$, a fact which is readily verified by direct substitution. This enables us to generate all the solutions χ_l, $l = 1, 2, 3, \dots$, from a knowledge of χ_0. Putting $l = 0$ in equation (8.7), one solves immediately for χ_0 as $\sin \rho$ or $\cos \rho$. The latter is unacceptable as it implies $R \to \infty$ as $r \to 0$, leading to a gradient discontinuity. Hence

$$R_0(r) = (\sin \alpha r)/\alpha r \tag{8.8}$$

Applying the above recurrence formula gives

$$R_1(r) = (\sin \alpha r)/\alpha^2 r^2 - (\cos \alpha r)/\alpha r \tag{8.9}$$

$$R_2(r) = (3/\alpha^3 r^3 - 1/\alpha r)\sin \alpha r - (3 \cos \alpha r)/\alpha^2 r^2 \tag{8.10}$$

and so on. Boundary conditions demand $R(a) = 0$, implying $\alpha a = n\pi$, $n = 1, \dots, 3, \dots$ for the $l = 0$ case (equation (8.8)). This yields energy levels $E = n^2\pi^2\hbar^2/2\mu a^2$, which are identical to the odd-parity one-dimensional energy levels for the square well (equation (2.11)). The full normalized wave function for the ground ($n = 1$, $l = 0$) state is $(2\pi a r^2)^{-1/2} \sin(\pi r/a)$.

For $l > 0$ the boundary condition $R(a) = 0$ leads to a more complicated equation for the energy eigenvalues. In all cases, however, these conditions approach either $\tan \alpha a = 0$ or $\cot \alpha a = 0$ for large E. In this approximation the energy levels coincide with the one-dimensional case.

8.3 HYDROGEN ATOM

This problem was treated in a simplified way in section 4.2. Putting $V(r) = -e^2/4\pi\varepsilon_0 r$, equation (8.5) becomes

$$\left[\frac{1}{\rho^2}\frac{d}{d\rho}\left(\rho^2\frac{d}{d\rho}\right) + \frac{\lambda}{\rho} - \frac{1}{4} - \frac{l(l+1)}{\rho^2}\right]R(\rho) = 0 \qquad (8.11)$$

where

$$\rho = \alpha r, \quad \alpha^2 = -8\mu E/\hbar^2, \quad \lambda = 2\mu e^2/4\pi\varepsilon_0\alpha h^2$$

Previously we examined this equation for $l = 0$ and found the energy levels (equation (4.28)). Here we note that the l-dependent term is negligible for large ρ, so the asymptotic form of u will still be $\exp(-\rho/2)$ (we reject $\exp(\rho/2)$ solutions which blow up as $\rho \to \infty$). Seeking solutions once again of the form $F(\rho)\exp(-\rho/2)$ we arrive at a modified version of equation (4.25)

$$\frac{d^2F}{d\rho^2} + \left(\frac{2}{\rho} - 1\right)\frac{dF}{d\rho} + \left[\frac{\lambda-1}{\rho} - \frac{l(l+1)}{\rho^2}\right]F = 0 \qquad (8.12)$$

Substitution of ρ^k in equation (8.12) shows the leading power to be ρ^{k-1}, with coefficients $\lambda - k - 1$ once again, so that the quantization condition (equation (4.27)) is totally unaffected by the presence of the l term. Thus, as stated in section 4.2, the energy levels (equation (4.28)) remain valid for all l. This l degeneracy is a special 'accidental' feature of the Coulomb-type potential and does not occur for other choices of $V(r)$.

To find the eigenfunctions we must solve equation (8.12) properly. To do this we put $F(\rho) = \Sigma_n a_n\rho^n$. Suppose the lowest power in this series is ρ^k. Substitution in equation (8.12) gives for the lowest-order term $[k(k+1) - l(l+1)]\rho^{k-2}$. This cannot cancel against a lower power contribution from F (there is none), so equation (8.12) requires $k(k+1) = l(l+1)$, implying that $k = l$ or $-(l+1)$. However, the latter choice is unacceptable as $\rho^{-(l+1)}$ blows up at $\rho = 0$. An acceptable series solution is, therefore, of the form

$$F(\rho) = \sum_{p=0} a_p\rho^{p+1} \qquad (8.13)$$

Substituting equation (8.13) into equation (8.12) we obtain from the coefficient of ρ^{p+l-1}

$$a_{p+1}(p+1)[p+2(l+1)] = a_p(p+l+1-\lambda) \qquad (8.14)$$

We see from (8.14) that for large p the ratio $a_{p+1}/a_p \to p^{-1}$. This is the same as for the series $\exp(\rho)$, which implies $u \sim \exp(\rho/2)$ – again unacceptable. Hence the series must terminate at some finite value of p, i.e. we admit only *polynomial* solutions F. This can be achieved by

$$\lambda - l - 1 = 0, 1, 2, \ldots \tag{8.15}$$

(note that λ must exceed l) as this will cause the coefficient of a_p in equation (8.15) to vanish at some value of p. In that case, $a_{p+1} = 0$ and all further terms, which are given recursively in terms of a_{p+1}, will also vanish. Noting that $l \geqslant 0$, equation (8.15) requires $\lambda = n$, $n = 1, 2, 3, \ldots$ $(l \leqslant n - 1)$ exactly as we found in section (4.2). This condition leads to the energy eigenvalues (equation (4.28))

$$E = -me^4/2(4\pi\varepsilon_0)^2\hbar^2 n^2 \tag{8.16}$$

Starting at $p = 0$, we can now use equation (8.14) to construct the functions $F(\rho)$ which are known to mathematicians as associated Laguerre polynomials. The radial eigenfunctions $R \equiv \exp(-\rho/2)F(\rho)$ thus carry two labels, n and l. The full hydrogenic wave functions are

$$u_{nlm}(r, \theta, \phi) = N_{nl} Y_{lm}(\theta, \phi) R_{nl}(r) \tag{8.17}$$

where the normalization constant is

$$N_{nl} = \left\{ \left(\frac{me^2}{2\pi\varepsilon_0 n} \right)^3 \frac{(n-l-1)!}{2n[(n+l)!]^3} \right\}^{1/2} \tag{8.18}$$

The Y_{lm} are already normalized. A few examples of u_{nlm} are given in Table 8.1. These expressions may also be used for one-electron ions with nuclear charge Z by replacing e^2 by Ze^2. Note that for $E > 0$ there are no bound states and the energy levels form a continuum.

It is of interest to obtain the probability $P_{nl}(r)\,dr$ of finding the electron at a distance between r and $r + dr$ from the nucleus. The

Table 8.1 *Some hydrogen atom stationary wave functions* u_{nlm}

$u_{100} = (\pi a_0^3)^{-1/2} \exp(-r/a_0)$

$u_{200} = (8\pi a_0^3)^{-1/2}(1 - r/2a_0)\exp(-r/2a_0)$

$u_{210} = (8\pi a_0^3)^{-1/2}(r/2a_0)\cos\theta\exp(-r/2a_0)$

$u_{21\pm1} = (\pi a_0^3)^{-1/2}(r/8a_0)\sin\theta\exp(\pm i\phi)\exp(-r/2a_0)$

$u_{300} = (27\pi a_0^3)^{-1/2}(1 - 2r/3a_0 + 2r^2/27a_0^2)\exp(-r/3a_0)$

$u_{310} = (2/27)(2/\pi a_0^3)^{1/2}(r/a_0)(1 - r/6a_0)\cos\theta\exp(-r/3a_0)$

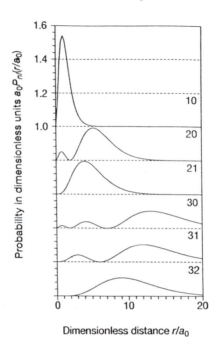

Fig. 8.1 This shows dimensionless plots of $P_{nl}a_0$ versus r/a_0. $P_{nl}(r)\,dr$ is the probability of finding the electron at a distance between r and $r+dr$ from the nucleus. Each curve of $P_{nl}a_0$ versus r/a_0 is labelled with two integers, representing the corresponding n and l values. Note the effect of the centrifugal force in repelling the wave function from the nucleus. Note also that the functions have $n-l-1$ zeroes in addition to those at $r=0$ and $r=\infty$.

answer is that

$$P_{nl}\,dr = |N_{nl}R_{nl}|^2 4\pi r^2\,dr \qquad (8.19)$$

Figure 8.1 shows dimensionless plots of $P_{nl}a_0$ versus r/a_0.

A number of small corrections to equation (8.16) have been studied both theoretically and experimentally. They arise from the effects of special relativity, the coupling between electron and nuclear magnetic moments and the electron's orbital magnetic moment, and finally some subtle, tiny, corrections due to quantum field theory.

The energy levels (equation 8.16)) depend on neither l nor m, so each level has a degeneracy of $2\Sigma_0^{n-1}(2l+1)=2n^2$, the factor 2 coming from the existence of two spin states. One effect of the coupling between the electron's spin and orbital magnetic moments is to modify the potential $V(r)$ from the Coulomb form and thus lift the 'accidental' l degeneracy. The result is that each energy level with $l > 0$ is split and a careful inspection of the spectra of hydrogen-like atoms reveals the presence of this 'fine structure' in the levels. In the next chapter, we consider some of these effects in more detail.

9

Some atomic physics

We know that even a single electron in a hydrogen atom generally has orbital angular momentum L (with $(2l + 1)$ possible orientations where l is a integer) and spin angular momentum S (with two possible orientations commonly described as spin up and spin down or $s = \pm 1/2$). In more complicated atoms there may be many components of angular momentum; for example, the helium atom has two electrons both of which have orbital and spin angular momenta. Also, in many atomic systems of physical interest we need to know how to combine angular momenta. This chapter explains how to do it.

9.1 ADDITION OF ANGULAR MOMENTA

Often, the notation adopted for a general angular momentum is J (including orbital and spin components) and the problem to be addressed might be that of the quantum mechanically allowed combinations of $J_1 + J_2$.

If we represent J in terms of its Cartesian coordinates

$$\hat{J} = (\hat{J}_x, \hat{J}_y, \hat{J}_z) \tag{9.1}$$

then we can follow exactly the logic associated with equations (7.33) to (7.44) inclusive. The change of notation from L to J is done here simply because it is usual to reserve L for orbital angular momentum, S for spin angular momentum, and J for total angular momentum. So for present purposes we denote the raising and lowering operators just as in equation (7.33) by

$$\hat{J}_\pm = \hat{J}_x \pm i\hat{J}_y \tag{9.2}$$

What these operators do is to convert a given eigenstate into the next one up or down the ladder of energies. We summarize the outcome by writing

$$\hat{J}^2 \hat{J}_\pm | j, m_j \rangle = \hbar^2 j(j+1) \hat{J}^\pm | j, m_j \rangle \tag{9.3}$$

and

$$\hat{J}_z \hat{J}_\pm | j, m_j \rangle = \hbar(m_j \pm 1) \hat{J}^\pm | j, m_j \rangle \tag{9.4}$$

where for convenience we use $| j, m_j \rangle$ to denote the simultaneous eigenstates of $\hat{J}^2$ and $\hat{J}_z$ with quantum numbers j and m_j. These results show that the raising and lowering operators do not alter j but they step m_j up or down by 1, except of course that upward steps cannot take place from the top of the ladder or downward steps from the bottom. The algebra given in equations (7.33)–(7.44) attends to these matters and finds that j must be positive (or zero) and may be integral or half-integral; m_j may take values in steps of unity in the range $-j$ to $+j$. For example with $j = 3/2$, m_j may only take one of the values $-3/2, -1/2, +1/2$ or $+3/2$. The number of allowed values of m_j for a given value of j is clearly $2j + 1$.

The angular momentum states for a composite system consisting of two subsystems can be characterized by either of two sets of good quantum numbers. The two representations are called the uncoupled and coupled. We discuss this next in the context of a precise example in which orbital and spin angular momenta are to be combined.

9.2   ADDITION OF ORBITAL AND SPIN ANGULAR MOMENTA

We consider here one of the simplest cases, in which the total angular momentum operator for an electron can be written as the sum of the orbital and spin angular momentum operators:

$$\hat{J} = \hat{L} + \hat{S} \tag{9.5}$$

where $\hat{J}$ has the components

$$\hat{J}_x = \hat{L}_x + \hat{S}_x; \quad \hat{J}_y = \hat{L}_y + \hat{S}_y; \quad \hat{J}_z = \hat{L}_z + \hat{S}_z \tag{9.6}$$

From the known commutation relations for $\hat{L}$ and $\hat{S}$ (section 7.1 and exercise 9.1), one can readily show that $\hat{J}^2$ and $\hat{J}_z$ commute, and it follows that

$$\hat{J}^2 | j, m_j \rangle = j(j+1)\hbar^2 | j, m_j \rangle \tag{9.7}$$

and

$$\hat{J}_z | j, m_j \rangle = m_j \hbar | j, m_j \rangle \tag{9.8}$$

where $j \geqslant |m_j|$ and is half-integral for electrons for which the spin quantum number $s = \frac{1}{2}$.

The square of the total angular momentum operator is given by

$$\hat{J}^2 = \hat{L}^2 + \hat{S}^2 + 2\hat{\boldsymbol{L}} \cdot \hat{\boldsymbol{S}} \tag{9.9}$$

Because $\hat{\boldsymbol{L}}$ and $\hat{\boldsymbol{S}}$ operate on the different variables of position and spin $\hat{\boldsymbol{L}}$ commutes with all three components of $\hat{\boldsymbol{S}}$, it readily follows that $\hat{J}^2$, $\hat{L}^2$, and $\hat{S}^2$ mutually commute. Furthermore $\hat{J}_z$ commutes with all three of them. Thus the four operators

$$\hat{J}^2, \hat{J}_z, \hat{L}^2, \hat{S}^2 \quad \text{(coupled representation)} \tag{9.10}$$

all commute with each other, so we may construct a state $|j, m_j\rangle$ that is a simultaneous eigenstate of all of them. However, there exists an alternative set of four operators

$$\hat{L}^2, \hat{L}_z, \hat{S}^2, \hat{S}_z \quad \text{(uncoupled representation)} \tag{9.11}$$

which also commute with each other, and therefore possess simultaneous eigenstates $|l, m_l, s, m_s\rangle$. We can express the former eigenstates as a linear combination of the latter as follows, taking $s = \frac{1}{2}$ for the electron:

$$|j, m_j\rangle = \alpha |l, m_l - \tfrac{1}{2}, \tfrac{1}{2}, +\tfrac{1}{2}\rangle + \beta |l, m_l + \tfrac{1}{2}, \tfrac{1}{2}, -\tfrac{1}{2}\rangle \tag{9.12}$$

This is already an eigenstate of $\hat{J}_z$ with eigenvalue m_j. The quantum number l belongs of course to the orbital angular momentum eigenstate. The coefficients α and β play a role like the c_n coefficients introduced in equation (6.2), and are to be determined. We shall make use of raising and lowering operators defined as in equation (7.33) where $\hat{L}_x$ and $\hat{L}_y$ are given in equations (7.5) and (7.6). The character of the eigenfunctions and the normalization factor given in equations (7.22) and (7.23) are also incorporated. Altogether we find

$$\hat{L}_{\pm} |l, m_l\rangle = \hbar (l \pm m_l + 1)(l \mp m_l)^{1/2} |l, m_l \pm 1\rangle \tag{9.13}$$

while

$$\hat{S}_{\pm} |\pm \tfrac{1}{2}\rangle = 0$$

and

$$\hat{S}_{\pm} |\mp \tfrac{1}{2}\rangle = \hbar |\pm \tfrac{1}{2}\rangle \tag{9.14}$$

Now we want to find α and β. There are two relationships, which can be found by operating on $|j, m_j\rangle$ with $\hat{J}^2$ in two ways. One is to equate the right-hand side of equation (9.7) with the right-hand side of equation (9.12) multiplied by $j(j+1)\hbar^2$. The other way is to observe that

$$\hat{J}^2 = \hat{L}^2 + \hat{S}^2 + 2\hat{L}\cdot\hat{S}$$
$$= \hat{L}^2 + \hat{S}^2 + 2\hat{L}_z\hat{S}_z + \hat{L}_+\hat{S}_- + \hat{L}_-\hat{S}_+ \qquad (9.15)$$

The result of operating on equation (9.12) with this is straightforward but generates a long string of algebra. The reader is invited to write out the relationships in full. The two relationships can be solved for j and for β/α. The further requirement of normalization, $\alpha^2 + \beta^2 = 1$, then achieves the desired result which is perhaps simpler than might be expected. One finds

$$|j = l \pm \tfrac{1}{2}, m_l\rangle = (2l+1)^{-1/2}[\pm(l+\tfrac{1}{2}\pm m_l)^{1/2}]|l, m_l - \tfrac{1}{2}, \tfrac{1}{2}, +\tfrac{1}{2}\rangle$$
$$+ (l + \tfrac{1}{2} \mp m_l)^{1/2}|l, m_l + \tfrac{1}{2}, \tfrac{1}{2}, -\tfrac{1}{2}\rangle \qquad (9.16)$$

The value of expressing the eigenstates in this way, which may seem to be an unnecessary complication at this stage, will become more apparent when we deal below with the magnetic coupling between L and S. Meanwhile it is worth mentioning that the multipliers α and β are examples of what are known more generally as *Clebsch–Gordan coefficients*. Their general significance can be indicated by another example. Suppose we are interested in a composite system of two electrons without regard to the spin. In the state $|l, m_l, l_1, l_2\rangle$ it is known that these electrons have respective angular momentum quantum numbers l_1 and l_2, and total angular momentum and z component quantum numbers l and m_l. The question then arises as to what can be said about the result of measurement of L_{1z} and L_{2z}, given that the initial state was $|l, m_l, l_1, l_2\rangle$. Specifically, what is the probability that at fixed L^2 and L_z the measurement will find one electron with $L_{1z} = m_{l1}\hbar$ and the other with $L_{2z} = m_{l2}\hbar$? The answer is provided by the squared modulus of the appropriate Clebsch–Gordan coefficient. Having seen the amount of algebra required for the calculation of such coefficients, one is grateful that tables of them exist so that it is not necessary to repeat the labour for each and every case.

A useful case for the purposes of graphical representation (Fig. 9.1) is that of the addition of an angular momentum L to a spin angular momentum S to give a total angular momentum J, for the case $s = \tfrac{1}{2}$ and $l > 0$. What happens is that L and S couple to give the resultant J and, in the presence of an applied magnetic field, the J vector precesses around the field direction. It is more accurate to picture the orbital and spin magnetic moments as linked (see the next section)

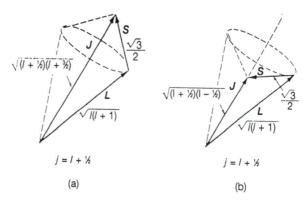

Fig. 9.1 This illustrates the addition of an orbital angular momentum L to a spin angular momentum S to give a total angular momentum J, for the case $s = \frac{1}{2}$, $l \neq 0$. The two cases are (a) S 'parallel' to L and (b) S 'anti-parallel' to L.

rather than precessing independently. This linkage, known as L–S or Russell–Saunders coupling, remains an accurate picture among lighter atoms where the electrons move much more slowly than light, even when several electrons are present. This scheme effectively ignores the interaction between electrons. (In heavy atoms, however, where electrons move faster, the inter-electron couplings must be taken into account.)

9.3 SPIN–ORBIT COUPLING

In sections 7.1 to 7.3 we described in quantum mechanical terms the orbital angular momentum of, for example, a single electron (of charge $-e$) bound to a nucleus (of charge Ze). We went on to introduce and treat electron spin in sections 7.4–7.6. Both forms of angular momentum produce magnetic moments μ_L and μ_S. The magnitudes are given in section 7.7, and we write them here in a slightly modified notation.

$$\mu_L = -\frac{g_L \mu_B}{\hbar} L \tag{9.17}$$

and

$$\mu_S = -\frac{g_S \mu_B}{\hbar} S \tag{9.18}$$

where μ_B is the Bohr magneton defined as $\mu_B \equiv e\hbar/2m_e$ (m_e being the mass of the electron), $g_L = 1$, and $g_S = 2$.

The two magnetic moments interact, and the mechanism can be visualized using the model of the atom as follows. Conventionally, the electron orbits the nucleus, but, from the electron's point of view, the nucleus orbits it just as from the our earthlings' point of view the sun appears to go round the Earth. In a frame of reference fixed on the electron, the nucleus moves in a circle around the electron and is seen by it as a current loop of radius r carrying a current $i = Zev/2\pi r$ where v is the orbital speed. The magnetic field at the electron would be

$$B = \frac{\mu_0 i}{2r} = \frac{\mu_0 Zev}{4\pi r^2} \qquad (9.19)$$

The direction of the magnetic field is (remaining with the Bohr model) perpendicular to the plane of the orbit and can be written vectorially as

$$B = -\frac{\mu_0 Ze}{4\pi} \frac{v \times r}{r^3} \qquad (9.20)$$

This can be written in another way which is preferred for some purposes. First we note that from classical mechanics the orbital angular momentum is $L = -m_e v \times r$. Second, we replace $-Ze^2/4\pi\varepsilon_0 r^2$ (the magnitude of the assumed Coulomb force on the electron) by $-dV(r)/dr$, where $V(r)$ is the potential energy of the electron. Third, we replace $\varepsilon_0\mu_0$ by $1/c^2$ where c is the velocity of light *in vacuo*. After these changes, equation (9.20) takes the form

$$B = \frac{1}{em_e c^2} \frac{1}{r} \frac{dV(r)}{dr} L \qquad (9.21)$$

Now the interaction energy between this field and the magnetic moment associated with electron spin, given by equation (9.18), would then classically be $\Delta E = (g_S\mu_B/\hbar)S \cdot B$. In view of the crudity of our model it would perhaps be surprising if this were exactly correct. In fact it is only out by a further factor of 2. (This is unrelated to the factor g_S/g_L also equal to 2.) It has to do with the fact that we have been working with a frame fixed on the electron, and a relativistic transformation of velocities back to the nuclear rest frame results in a reduction of the orientational potential energy by a factor of 2. Thus

$$\Delta E = \frac{g_S\mu_B}{2\hbar} B \cdot S \qquad (9.22)$$

$$= \frac{g_S \mu_B}{2em_e c^2 \hbar} \frac{1}{r} \frac{dV(r)}{dr} \mathbf{L} \cdot \mathbf{S} \qquad (9.23)$$

$$= \frac{1}{2m_e^2 c^2} \frac{1}{r} \frac{dV(r)}{dr} \mathbf{L} \cdot \mathbf{S} \qquad (9.24)$$

A bonus to this relatively simple result is that, although we have obtained it in a very unsafe way, using doubtful assumptions about the Bohr model, equation (9.24) can be shown to have general validity. In other words, it remains correct in a full theory of relativistic quantum mechanics. It is also valid for a general central potential $V(r)$, not merely for the Coulomb case.

Using equation (9.9) to substitute for $\mathbf{L} \cdot \mathbf{S}$ in equation (9.24) enables us to construct the Hamiltonian operator which includes the perturbation due to spin–orbit coupling. It is

$$\hat{H} = \hat{H}_0 + \hat{L}^2/2m_e r^2 + f(r)(\hat{J}^2 - \hat{L}^2 - \hat{S}^2) \qquad (9.25)$$

where

$$f(r) = \frac{1}{4m_e^2 c^2} \frac{1}{r} \frac{dV(r)}{dr} \qquad (9.26)$$

Even with the inclusion of the interaction term, the radial and angular parts of the time-independent Schrödinger equation can still be separated in the manner shown in section 8.1. The resulting perturbation in the Hamiltonian can be written as

$$\hat{H}' = f(r)\hbar^2 [j(j+1) - l(l+1) - s(s+1)] \qquad (9.27)$$

It readily follows from perturbation theory (Chapter 10) that an energy level with quantum numbers n, j, l, s will be shifted by

$$\langle f(r) \rangle \hbar^2 [j(j+1) - l(l+1) - s(s+1)] \qquad (9.28)$$

where $\langle f(r) \rangle$ is the expectation value of $f(r)$ calculated using the radial part of the unperturbed wavefunction. For the hydrogen atom this is easy to calculate using the radial wavefunctions $R_{nl}(r)$ discussed in section 8.3. The result of this calculation is that, for $l = 0$, $\langle f(r) \rangle$ is zero, while, for $l > 0$, it is given by

$$\langle f(r) \rangle \hbar^2 = \frac{1}{4} m_e c^2 \alpha^4 \frac{1}{n^3 l(l + \frac{1}{2})(l+1)} \qquad (9.29)$$

where $\alpha \equiv e^2/4\pi\varepsilon_0 \hbar c$ is the so-called fine structure constant whose numerical value is close to 1/137. Combining equations (9.28) and

(9.29) thus produces the result that for the hydrogen atom the energy shift is

$$\frac{1}{4} m_e c^2 \alpha^4 \frac{j(j+1) - l(l+1) - s(s+1)}{n^3 l(l + \frac{1}{2})(l+1)} \tag{9.30}$$

Taking the possible values of j as $l + \frac{1}{2}$ and $l - \frac{1}{2}$ (with $s = \frac{1}{2}$), it follows that levels with $l > 0$ are thus split into doublets. If we use the notation δE_{nl} to denote the splitting attributable to the spin–orbit interaction we obtain

$$\delta E_{nl} = \frac{1}{4} m_e c^2 \alpha^4 \frac{1}{n^3} \frac{1}{l(l+1)} \tag{9.31}$$

of which the upper has been shifted *up* by $[l/(2l+1)]\delta E_{nl}$ while the lower has been shifted *down* by $[(l+1)/(2l+1)]\delta E_{nl}$.

9.4 A FURTHER RELATIVISTIC EFFECT

Equation (9.31) gives an energy splitting which is proportional to α^4 which ensures that splittings due to spin–orbit interaction are small compared with the separations between energy levels which are proportional to α^2. There is another effect which is also proportional to α^4. This has to do with the fact that the kinetic energy of the electron requires a small relativistic correction. We shall not give the calculation here but the result is that when both effects are included the original energy level is shifted by

$$\frac{1}{4} m_e c^2 \alpha^4 \frac{1}{n^3} \left(\frac{1}{j + \frac{1}{2}} - \frac{3}{4n} \right) \tag{9.32}$$

for both values $l = j \pm \frac{1}{2}$.

Curiously, if one chooses a value of l (say, 3) and finds the two values of j (5/2 and 7/2) then the splitting calculated from equation (9.32) is exactly the same as the splitting calculated for the spin–orbit interaction alone (equation (9.31)). For the special case of $l = 0$, there is only one value of $j = 1/2$ and equation (9.32) gives the correct result of zero splitting. Some results are illustrated in Fig. 9.2 for the case $n = 4$, $l = 0, 1, 2, 3$. This type of splitting is indeed observed in the spectrum of hydrogen. Another example is that of the two familiar yellow D-lines of sodium vapour, one of which is split in two while the other is unaffected by spin–orbit coupling and therefore remains single. The sodium atom has one valence electron outside a set of closed shells. The potential experienced by that electron is not of a

Fig. 9.2 For the case of $n = 4$, there are four possible values of l $(0, 1, 2, 3)$. For $l = 0$ there is only one value of j, i.e. $\frac{1}{2}$. For each of the non-zero values of l there are two values of j equal to $l \pm \frac{1}{2}$. For each value of l we show the original, unshifted energy (plotted vertically in units of $m_e c^2 \alpha^4 / 256$) and the reduction for each corresponding j value caused by the combined effects of spin–orbit coupling and relativistic kinetic energy.

simple Coulomb form because of the screening by the closed shells, but the formalism is much the same as for hydrogen.

9.5 EFFECT ON A ONE-ELECTRON ATOM OF AN EXTERNALLY APPLIED MAGNETIC FIELD: ZEEMAN EFFECT

In this section we consider the shifts in energy levels resulting from a constant uniform applied external magnetic field $B_{ext} > 0$. In order to describe the effects it is necessary to add an extra term to equation (9.25) and in general this term has two parts, one (the linear part) proportional to B_{ext} and the other (the quadratic part) proportional to B_{ext}^2. The ratio of quadratic to linear parts is approximately $(ea_0^2/4\hbar)B_{ext} \approx 10^{-6} B_{ext}$. Thus for field magnitudes normally available in the laboratory (rarely in excess of 1 kT) the quadratic part of the extra term is safely neglected. With this proviso, and for fields B_{ext} applied along the z axis, we see from equations (9.17) and (9.18) that the extra term is

$$\frac{\mu_B}{\hbar}(g_L L_z + g_S S_z)B_{ext} \tag{9.33}$$

or, substituting the values $g_L = 1$ and $g_S = 2$,

$$\frac{\mu_B}{\hbar}(L_z + 2S_z)B_{ext} \tag{9.34}$$

With the addition of this term the Hamiltonian operator becomes

$$\hat{H} = \hat{H}_0 + \frac{\hat{L}^2}{2m_e r^2} + f(r)(\hat{J}^2 - \hat{L}^2 - \hat{S}^2) + \frac{\mu_B}{\hbar}(\hat{L}_z + 2\hat{S}_z)B_{ext} \quad (9.35)$$

The additional, fourth, term is known as the Zeeman term because its effects were first observed in 1896 by P. Zeeman. The third term describes the spin–orbit coupling energy derived in section 9.3. The time-independent Schrödinger equation with the above Hamiltonian can be solved numerically and the energy eigenvalues found as functions of B_{ext}. Figure 9.3 shows the result diagrammatically.

In section 8.3 we discussed the hydrogen atom in the absence of either spin–orbit or Zeeman terms in the Hamiltonian. In such a simplified view, the allowed energies given in equation (8.16) depend only upon the quantum number n. Consider the case $n = 2, l = 1$; the allowed values of j are 3/2 (orbital and spin angular momenta 'parallel') and 1/2 (orbital and spin angular momenta 'antiparallel'). $j = 3/2$ may have $m_j = 3/2, 1/2, -1/2$, or $-3/2$ (four possibilities) while $j = 1/2$ may have $m_j = +1/2$ or $-1/2$ (two possibilities). All six states are degenerate in the absence of spin–orbit and Zeeman effects. Spin–orbit coupling alone splits the level into a doublet, and

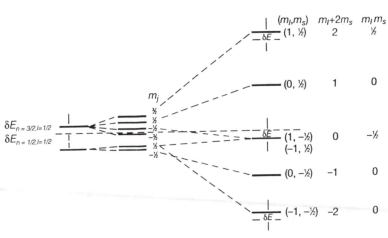

Fig. 9.3 The energies of the states with $l = 1$ in a one-electron atom (such as hydrogen) with a zero (left), weak (centre) and strong (right) magnetic field applied. In this context, strong does not mean arbitrarily strong; it means strong enough to produce shifts which are much larger than spin–orbit effects labelled δE, but not so large as to require the presence of quadratic terms in the Hamiltonian.

according to equation (9.28) the levels with $j = 3/2$ are shifted up by $\langle f(r)\rangle \hbar^2$ while those with $j = 1/2$ are shifted down by $2\langle f(r)\rangle \hbar^2$. The splitting is thus $3\langle f(r)\rangle \hbar^2$. Now for the hydrogen atom $\langle f(r)\rangle$ is readily calculated. With $n = 2$ and $l = 1$ Fig. 9.3 shows what happens to the energies associated with those combinations as B_{ext} is raised from zero. All six cases are affected in different ways by B_{ext} depending on the values of the various quantum numbers as listed in the figure. Because the energy shifts can be seen to cross over from the $B_{ext} = 0$ case to the high B_{ext} case it is common to divide the results into two or three ranges. We sketch the approaches below, but it should be borne in mind that they all follow from the same Hamiltonian given in equation (9.35).

1. *The strong-field Zeeman effect*

   In this case the magnetic field is sufficiently strong that the spin–orbit term can be neglected. It cannot be arbitrarily strong, however, because we still intend to neglect the quadratic part of the Zeeman term (this is perfectly justifiable approximation for laboratory experiments though possibly not for certain astro-physical situations). In this approximation, $\hat{H}$ commutes with $\hat{L}^2$, $\hat{S}^2$, $\hat{L}_z$, and $\hat{S}_z$ so that the corresponding eigenvalues are constants of the motion and l, s, m_l, and m_s are all good quantum numbers. The energy levels are then simply

   $$E_{nm_lm_s} = E_n + \mu_B B_{ext}(m_l + 2m_s) \qquad (9.36)$$

   where E_n is given by equation (8.16). In our example we take $l = 1$ and there are five possible values of $m_l + 2m_s$, equal to 2, 1, 0, -1, -2, and Fig. 9.3 shows these as equally spaced by an amount $\mu_B B_{ext}$.

2. *The Paschen–Back effect*

   In this case the spin–orbit term is treated as a small but not negligible perturbation. Use of first-order perturbation theory (section 10.1) to estimate the shift of energy levels from the strong-field case above gives the result

   $$\delta E \equiv E_{nlm_lm_s} - E_{nm_lm_s} = \langle f(r)\rangle m_l m_s \hbar^2 \qquad (9.37)$$

   This can conveniently be written in terms of the spin–orbit splitting energy given in equation (9.31). Then

   $$\delta E = \delta E_{nl} \frac{m_l m_s}{(l + \frac{1}{2})} \qquad (9.38)$$

This term must be added to that of equation (9.36) and so

$$E_{nm_lm_s} = E_n + \mu_B B_{ext}(m_l + 2m_s) + \delta E_{nl}\frac{m_lm_s}{(l+\frac{1}{2})} \qquad (9.39)$$

3. *The weak-field Zeeman effect*

In this case we consider what happens to the energy levels which are degenerate when the external field is strictly zero but which are split by the least external field. A proper calculation of this is too complicated for inclusion here. There is, however, a non-rigorous but physical argument which arrives at the correct answer. This goes as follows. If we assume the field to be so weak that the eigenstates are unchanged from the zero-field case, then we can picture the J vector (i.e. $L + S$ rather than L or S separately) as retaining its integrity with $2j + 1$ states each denoted by a particular value of m_j. If we also assume that components of the magnetic moment in directions other than that of the total angular momentum will have no effect on the interaction energy, then we are interested in the component of the magnetic moment parallel to the unit vector J/J. This component is $(J/J)\cdot(e/2m_e)(L + 2S)$ with $J = L + S$. Multiplication of this component of magnetic moment by the small applied field B_{ext} and some simple algebra reduces this to the result

$$(eB_{ext}J_z/2m_e)[1 + (J^2 - L^2 + S^2)/2J^2]$$

We can now choose to write this as

$$\delta E = g\mu_B m_j B_{ext} \qquad (9.40)$$

where

$$g = 1 + \frac{j(j+1) - l(l+1) + s(s+1)}{2j(j+1)} \qquad (9.41)$$

is known as the Landé g factor. It is of interest to note that if $s = 0$ then $g = g_L = 1$, and if $l = 0$ then $g = g_S = 2$, two limiting cases (pure orbit or pure spin) which could be expected on the grounds that they both correspond to the magnetic moment being parallel to the angular momentum.

10

Approximation methods

As in classical mechanics the number of exactly soluble quantum mechanical problems is very limited and in most cases of practical interest numerical techniques and approximation methods are unavoidable. The availability of fast computers has turned quantum computation into a major industry. In this chapter we shall look at two approximation techniques.

10.1 PERTURBATION THEORY

In many cases the system of interest differs from an exactly soluble system by only a small disturbance, enabling an approximation to be made by expanding in powers of a smallness parameter. We treat here the theory for stationary states only and extend to time-dependent systems in Chapter 11.

Suppose that the undisturbed system has Hamiltonian $\hat{H}_0$, energy eigenfunctions $u_n^{(0)}$ and eigenvalues $E_n^{(0)}$:

$$\hat{H}_0 u_n^{(0)} = E_n^{(0)} u_n^{(0)} \tag{10.1}$$

The disturbance adds a small extra piece $\hat{H}'$ to the Hamiltonian:

$$\hat{H} = \hat{H}_0 + \hat{H}' \tag{10.2}$$

In seeking to solve the Schrödinger equation with this perturbed Hamiltonian, it is convenient to work with series expansions by introducing a dimensionless parameter λ. The value of this parameter does not come into the results, but it may aid understanding to constrain it within the range 0 to 1. We concentrate now on the equation

$$\hat{H} = \hat{H}_0 + \lambda \hat{H}' \tag{10.3}$$

The aim is to obtain expressions for the disturbed energy eigenvalues and eigenfunctions in terms of power series in λ. Since $\lambda \leqslant 1$ and $\hat{H}' \leqslant \hat{H}_0$ it is a reasonable supposition (which we shall not attempt to prove here) that these series are convergent. The series are written as

$$E_n = E_n^{(0)} + \lambda E_n^{(1)} + \lambda^2 E_n^{(2)} + \cdots \tag{10.4}$$

and

$$u_n = u_n^{(0)} + \lambda u_n^{(1)} + \lambda^2 u_n^{(2)} + \cdots \tag{10.5}$$

where the subscript n labels the eigenstate while the bracketed numerical superscript labels the terms in the expansions. Now the Schrödinger equation for the perturbed system,

$$(\hat{H} - E_n)u_n = 0 \tag{10.6}$$

must be true order by order in powers of λ if equations (10.4) and (10.5) are to be true for arbitrary (small) λ. Substituting equations (10.3), (10.4) and (10.5) into equation (10.6) and equating the coefficient of each power of λ to zero gives an infinite series of equations, the first (zeroth order) being equation (10.1). The second (first order) gives

$$E_n^{(1)}u_n^{(0)} = \hat{H}'u_n^{(0)} + (\hat{H}_0 - E_n^{(0)})u_n^{(1)} \tag{10.7}$$

Now the eigenfunctions $u_n^{(0)}$ form a complete orthonormal set so we may expand any function, including the correction function $u_n^{(1)}$, as follows

$$u_n^{(1)} = \sum_m a_{nm}u_m^{(0)} \tag{10.8}$$

with some coefficients a_{nm}.

Using this form in equation (10.7) we get

$$E_n^{(1)}u_n^{(0)} = \hat{H}'u_n^{(0)} + \sum_m a_{nm}(E_m^{(0)} - E_n^{(0)})u_m^{(0)} \tag{10.9}$$

Multiplying from the left by $u_n^{(0)*}$ and integrating, the final term vanishes for $n \neq m$ because the $u_n^{(0)}$ are orthogonal while the $m = n$ term vanishes because of the $E_m^{(0)} - E_n^{(0)}$ factor, reducing equation (10.9) to

$$E_n^{(1)} = H'_{nn} \equiv \int u_n^{(0)*}\hat{H}'u_n^{(0)}\,d\tau \tag{10.10}$$

where the normalization condition has been used on the left-hand side and H'_{nm} are the diagonal elements of a matrix $H'_{mn} \equiv \langle u_m^{(0)} | \hat{H}' | u_n^{(0)} \rangle$ (see section 6.2). So, knowing the exact unperturbed eigenfunctions $u_n^{(0)}$ and given the perturbation Hamiltonian $\hat{H}'$, the first-order corrections to the energy levels are simply the expectation values of $\hat{H}'$ in the unperturbed states $u_n^{(0)}$.

The corrected eigenfunctions follows from equation (10.9) Multiplying by $u_m^{(0)*}$ and integrating gives

$$a_{nm} = \frac{H'_{mn}}{E_n^{(0)} - E_m^{(0)}} \qquad \text{if } n \neq m \qquad (10.11)$$

The value of a_{nn} must be determined by normalizing u_n. When this is done it turns out to be purely imaginary and can be made to vanish by a suitable choice of unobservable phase. Putting equation (10.11) into equation (10.8) we may rewrite equation (10.5):

$$u_n = u_n^{(0)} + \sum_{m \neq n} \frac{H'_{mn} u_m^{(0)}}{E_n^{(0)} - E_m^{(0)}} + \cdots \qquad (10.12)$$

where we are using equation (10.5) with $\lambda = 1$ (within its allowed range) so as to arrive at the solution of equation (10.2). The corresponding first-order energy similarly follows from equation' (10.4), again with $\lambda = 1$.

If the first-order approximation is inadequate we can continue the series to order λ^2 to obtain

$$E_n^{(2)} = \sum_{m \neq n} |H'_{mn}|^2 (E_n^{(0)} - E_m^{(0)}) \qquad (10.13)$$

In the foregoing we have assumed that the unperturbed eigenstates are non-degenerate. If two or more values of $E_n^{(0)}$ are the same, equation (10.12) contains infinite terms. The trouble comes from the fact that degenerate eigenfunctions are not unique – any linear combination is also an eigenfunction with the same eigenvalues (see section 6.2) – whereas if the perturbation lifts the degeneracy the perturbation eigenfunctions are unique. If in equation (10.8) we expand using an arbitrary set of degenerate eigenfunctions, there is obviously a likelihood of a mismatch with the u_n. The infinite terms in equation (10.12) reflect the discontinuous jump that occurs as soon as the perturbation is switched on. The problem is easily cured by constructing the correct linear combination of degenerate $u_n^{(0)}$ to match smoothly onto the u_n. This is accomplished by selecting that linear combination that diagonalizes the degenerate submatrix of the matrix H'_{mn}.

10.2 APPLICATIONS OF PERTURBATION THEORY

In this section we shall illustrate the foregoing results using a number of well-known examples.

10.2.1 The anharmonic oscillator

A simple harmonic oscillator is disturbed by an additional term αx^4 in $V(x)$, where α is a constant. Calculate the first-order correction to the ground state energy.

The ground state wave function is given by equation (4.10). From equation (10.10) we have

$$E_0^{(1)} = \alpha \left(\frac{m\omega}{\pi\hbar}\right)^{1/2} \int_{-\infty}^{\infty} x^4 e^{-m\omega x^2/\hbar} \, dx$$

$$= 3\hbar^2 \alpha/4m^2\omega^2$$

10.2.2 Electric polarizability of hydrogen atom

An electric field E is applied to a hydrogen atom in its ground state. Calculate the energy shift to second order.

As the ground state is spherically symmetric, we may choose E to lie along the z axis, so $H' = -eEz$. From equation (10.10) we see that the integrand is an odd function (changes sign under reflection in the origin). This is because $|u_n^{(0)}|^2$, being spherically symmetric, is even, whereas z is odd. Hence when the integration is performed over all space, cancellation occurs to give a zero result. We therefore proceed to second order. From equation (10.13)

$$E_0^{(2)} = -e^2 E^2 \sum_{m \neq 0} |\langle u_m|z|u_0\rangle|^2/(E_m^{(0)} - E_0^{(0)})$$

The summation is very complicated. The crucial feature, however, is that the shift is proportional to E^2. Compare this with the classical result of $-\frac{1}{2}\alpha E^2$ for a system of electric polarizability α. The electric field first induces a dipole moment equal to αE, which then couples to the field to give an energy $-\frac{1}{2}\alpha E^2$. Only if the system already has a dipole moment would the shift be proportional to E. This is the case for the excited states of the hydrogen atom.

Rather than evaluating the above summation for E_0, we can actually solve equation (10.7) exactly in this case, to obtain E_0 directly, without the need for the eigenfunction expansion (equation

(10.8)). This can be done because $E_0^{(1)} = 0$ for this problem. Thus equation (10.7) reduces to

$$\left(-\frac{\hbar^2}{2m}\nabla^2 - \frac{e^2}{4\pi\varepsilon_0 r} + \frac{me^4}{32\pi^2\varepsilon_0^2\hbar^2}\right)u_n^{(1)} - eEzu_n^{(0)} = 0$$

which has the solution

$$u_n^{(1)} = (4\pi\varepsilon_0/e^2)eEz(\tfrac{1}{2}r + a_0)u_n^{(0)}$$

The second-order energy correction is given by $\langle u_n^{(1)}|\hat{H}'|u_n^{(0)}\rangle$ (it can be shown), which is readily evaluated:

$$E_0^{(2)} = -\tfrac{9}{4}e^2E^2(4\pi\varepsilon_0/e^2)^4(\hbar^2/m)^3$$

10.2.3 Electrons in a metal

Using the periodic one-dimensional potential $V(x) = V_0\cos(2\pi x/a)$ as a model of a crystalline solid, show that there is no first-order energy shift unless the electron's momentum $p = \pm\hbar\pi/a$.

In the absence of the perturbing potential the electrons are free, with wave functions as in equation (4.35) and energy $E = \hbar^2k^2/2m$. We find

$$H'_{kk} = L^{-1}\int_0^L V_0\cos(2\pi x/a)\,dx = 0$$

where $L = Na$ is the length of a lattice of N ions of spacing a. Similarly $H'_{k-k} = 0$. Hence the lattice leaves the electron's energy undisturbed for these cases.

Now the states $|k\rangle$ and $|-k\rangle$ actually have the same energy $\hbar^2k^2/2m$, so there is a two-fold degeneracy involved. As explained at the end of section 10.1, the results of perturbation theory can still be used but only if the degenerate submatrix is first diagonalized. We must therefore also consider the off-diagonal matrix elements

$$H'_{k-k} = H'^{*}_{-kk} = L^{-1}\int_0^L e^{-2ikx}V_0\cos(2\pi x/a)\,dx$$

$$= V_0/2 \qquad\qquad \text{if } k = \pm\pi/a$$

$$= 0 \qquad\qquad \text{otherwise}$$

Thus only if $p \equiv \hbar k = \pm\hbar\pi/a$ does the perturbing potential shift the energy. Diagonalization is accomplished by solving the determinantal equation

$$\begin{vmatrix} H'_{kk} - E^1 & H'_{k-k} \\ H'_{-kk} & H'_{-k-k} - E^{(1)} \end{vmatrix} = \begin{vmatrix} -E^{(1)} & V_0/2 \\ V_0/2 & -E^{(1)} \end{vmatrix} = 0$$

which has solutions $E^{(1)} = \pm V_0/2$. Thus the degenerate levels are split apart by V_0.

10.2.4 The helium atom

The helium atom consists of two electrons in orbit around a nucleus of charge $+2e$. In the absence of interactions between the electrons each would have a hydrogen-like wave function. We may treat the inter-electron coupling as a perturbation.

From the discussion in 4.6 we must ensure that the total wave function Ψ is antisymmetric. Taking into account the spin eigenstates, a possible unperturbed state is $u_1(r_1)u_2(r_2)|\uparrow_1\rangle|\downarrow_2\rangle$. We could also have states with 1 and 2 interchanged on the u functions, or the spin states, or we could have both spins either up or down. That gives eight combinations in all, which may be combined to give four antisymmetric wave functions

$$\left.\begin{aligned} &\frac{1}{\sqrt{2}}[u_1(r_1)u_2(r_2) - u_2(r_1)u_1(r_2)]|\uparrow_1\rangle|\uparrow_2\rangle \\[2mm] &\frac{1}{\sqrt{2}}[u_1(r_1)u_2(r_2) - u_2(r_1)u_1(r_2)]|\downarrow_1\rangle|\downarrow_2\rangle \\[2mm] &\frac{1}{2}[u_1(r_1)u_2(r_2) - u_2(r_1)u_1(r_2)][|\uparrow_1\rangle|\downarrow_2\rangle + |\uparrow_2\rangle|\downarrow_1\rangle] \end{aligned}\right\} \quad (10.14)$$

$$\frac{1}{2}[u_1(r_1)u_2(r_2) + u_2(r_1)u_1(r_2)][|\uparrow_1\rangle|\downarrow_2\rangle - |\uparrow_2\rangle|\downarrow_1\rangle] \quad (10.15)$$

In equations (10.14) the two electron spins are aligned, giving a total spin of $\hbar$, with z component $+\hbar$, $-\hbar$ and 0 respectively. In equation (10.15) the space part of the wave function is symmetric but the spin part is antisymmetric. Here the spins are opposed to give a total spin of zero (recall the discussion at the end of section 7.8).

The electrostatic force between the electrons introduces the perturbation

$$\hat{H}' = e^2/4\pi\varepsilon_0 r_{12} \quad (10.16)$$

where $r_{12} = |r_1 - r_2|$. Although all four states shown in (10.14) and (10.15) are degenerate, the relevant submatrix is, happily, already diagonal, so we may proceed and use the results of non-degenerate perturbation theory. They perturbation (equation (10.16)) does not affect the spins so the states given by (10.14) remain a degenerate triplet. There is an energy gap, however, between this triplet and the spin zero singlet (given by (10.15)) on account of the different space part of the wave functions. The first-order shifts $E_1^{(1)}$, $E_2^{(1)}$ in the triplet and singlet states respectively are, from equation (10.10),

$$E_1^{(1)} = H'_{11} = I_1 - I_2 \tag{10.17}$$

$$E_2^{(1)} = H'_{22} = I_1 + I_2 \tag{10.18}$$

$$I_1 = \frac{e^2}{4\pi\varepsilon_0} \int \int u_1^*(r_1) u_2^*(r_2) \frac{1}{r_{12}} u_1(r_1) u_2(r_2) \, d\tau_1 \, d\tau_2 \tag{10.19}$$

$$I_2 = \frac{e^2}{4\pi\varepsilon_0} \int \int u_1^*(r_1) u_2^*(r_2) \frac{1}{r_{12}} u_1(r_2) u_2(r_1) \, d\tau_1 \, d\tau_2 \tag{10.20}$$

The integrand of I_1 has the form $\rho(r_1)\rho(r_2)/r_{12}$, where ρ is an electric charge density, so it is like a Coulomb energy for two classical interpenetrating clouds of charge. On the other hand, I_2 has no classical analogue; it is referred to as the 'exchange' integral. The evaluation of the integrals is complicated but straightforward. One finds both I_1 and $I_2 > 0$ so that the singlet state has higher energy than the triplet ($E_2^{(1)} > E_1^{(1)}$), implying that when the spins are parallel the particles tend to keep their distance and so suffer less electrostatic repulsion. This is the Pauli principle at work. In the case of the ground state both electrons are in the lowest energy configuration, corresponding to the ψ_{100} hydrogenic wave functions. They thus have the same wave function, u, which causes the states (10.14) to vanish, i.e. the triplet state does not exist. The Pauli principle therefore requires the ground state to be a singlet, in spite of the fact that a triplet state would have had lower energy.

10.3 THE VARIATIONAL METHOD

This provides a very simple method of approximating energy eigenvalues. It is most useful for computing ground state energies and its success depends on our being able to make a reasonable guess for the form of the ground state wave function.

Suppose we guess a (normalized) trial wave function u_t for the ground state and compute the expectation value of the Hamiltonian

$$\langle u_t | \hat{H} | u_t \rangle = \int u_t^* \hat{H} u_t \, d\tau \tag{10.21}$$

We shall show that this value can never be lower than the true ground state energy. To see this we expand u_t in terms of the actual eigenfunctions u_n of $\hat{H}$:

$$u_t = \sum_n a_n u_n \tag{10.22}$$

where normalization of u_t requires

$$\sum_n |a_n|^2 = 1 \tag{10.23}$$

Using equation (10.22) in equation (10.21) we get

$$\langle \hat{H} \rangle = \int \left(\sum_n a_n^* u_n^* \right) \hat{H} \left(\sum_m a_m u_m \right) d\tau$$

$$= \sum_n \sum_m a_n^* a_m \int u_n^* \hat{H} u_m \, d\tau$$

$$= \sum_n \sum_m a_n^* a_m E_m \delta_{nm}$$

$$= \sum_n |a_n|^2 E_n \tag{10.24}$$

As $E_n \geqslant E_0$ (the ground state energy) we can certainly say from equation (10.24)

$$\langle \hat{H} \rangle \geqslant \sum_n |a_n|^2 E_0 = E_0 \sum_n |a_n|^2 \tag{10.25}$$

So using equation (10.23) we get from equation (10.25)

$$\langle u_t | \hat{H} | u_t \rangle \geqslant E_0 \tag{10.26}$$

The strategy is thus to make a plausible guess at u_t and to include some adjustable parameters. The values of these parameters can then be varied until $\langle \hat{H} \rangle$ is minimized. This is then the best fit to E_0.

10.3.1 Using the trial function $\exp(-\alpha x^2)$ estimate the ground state of the simple harmonic oscillator

This is a favourite because it actually gives the correct answer exactly. The factor α is our adjustable parameter to be varied.

First we normalize u_t using $\int \exp(-2\alpha x^2)\,dx = \sqrt{(\pi/2\alpha)}$. This gives $u_t = (2\alpha/\pi)^{1/4}\exp(-\alpha x^2)$. Then we refer to equation (4.1) for $\hat{H}$ and compute

$$\langle u_t|\hat{H}|u_t\rangle = \left(\frac{2\alpha}{\pi}\right)^{1/2}\int_{-\infty}^{\infty} e^{-\alpha x^2}\left(-\frac{\hbar^2}{2m}\frac{d^2}{dx^2}+\frac{1}{2}Kx^2\right)e^{-\alpha x^2}\,dx$$

$$= \hbar^2\alpha/2m + m\omega^2/8\alpha \tag{10.27}$$

We may now minimize equation (10.27) with respect to α. We find $d\langle\hat{H}\rangle/d\alpha = 0$ when $\alpha = m\omega/2\hbar$, which gives, in fact, the correct ground state wave function (equation (4.10)). The corresponding energy is, from equation (10.27)

$$\langle\hat{H}\rangle = \tfrac{1}{2}\hbar\omega$$

By way of comparison let us pick a less judicious choice of trial function,

$$u_t = \alpha^{-2}(15/16\alpha)^{1/2}(\alpha^2 - x^2) \qquad |x| < \alpha$$
$$= 0 \qquad |x| > \alpha$$

This inverted parabola is a bad attempt to model the Gaussian function, equation (4.10), but has the virtue that the integrals are easy to evaluate. Let us see how well it does.

$$\langle\hat{H}\rangle = \frac{15}{16\alpha^5}\int_{-\alpha}^{\alpha}(\alpha^2 - x^2)\left(-\frac{\hbar^2}{2m}\frac{d^2}{dx^2}+\frac{1}{2}Kx^2\right)(\alpha^2 - x^2)\,dx$$

$$= 5\hbar^2/4m\alpha^2 + m\omega^2\alpha^2/14$$

Finally, $\langle\hat{H}\rangle$ is minimized if $\alpha = (35\hbar^2/2m^2\omega^2)^{1/4}$, whereupon $\langle\hat{H}\rangle = 0.598\hbar\omega$, which is a reasonable approximation.

11

Transitions

Much of the theory presented so far has been directed towards finding eigenvalues and eigenfunctions for stationary states. In many practical cases a stationary system is perturbed in some way, inducing it to undergo *transitions* between the otherwise stationary states. Perhaps the best example concerns the radiative transitions of atomic electrons in which the disturbance of an external electromagnetic field causes the electrons to jump between energy levels, emitting or absorbing photons. Along with many similar processes, it may be dealt with by the use of time-dependent perturbation theory.

11.1 TIME-DEPENDENT PERTURBATION THEORY

Suppose a system with a Hamiltonian $\hat{H}_0(r)$ has a known set of stationary states

$$\hat{H}_0 u_n = E_n u_n \tag{11.1}$$

and is perturbed by a time-dependent disturbance described by $\hat{H}'(r, t)$. We want to find how the wave function $\psi(r, t)$ of the disturbed system evolves with time. The Schrödinger equation contains this information:

$$\hat{H}\psi(r, t) \equiv (\hat{H}_0 + \hat{H}')\psi = i\hbar \frac{\partial \psi(r, t)}{\partial t} \tag{11.2}$$

The stationary state wave functions of $\hat{H}_0$ form a complete orthonormal set, so we may expand ψ:

$$\psi(r, t) = \sum_n a_n(t) u_n(r) e^{-iE_n t/\hbar} \tag{11.3}$$

where the expansion coefficients a_n are now time dependent to take into account the time dependence of $\hat{H}'$, and we have included explicitly the usual $\exp(-iE_n t/\hbar)$ factor associated with the stationary state wave functions. Substituting equation (11.3) into equation (11.2) we get (in what follows a dot denotes differentiation with respect to time)

$$\sum_n a_n(\hat{H}_0 + \hat{H}')u_n\, e^{-iE_n t/\hbar} = \sum_n (a_n E_n + i\hbar\dot{a}_n)u_n\, e^{-iE_n t/\hbar}$$

The first terms on the left and right cancel using equation (11.1) Multiplying by u_m^* and integrating, the remainder gives

$$\sum_n a_n\left(\int u_m^* \hat{H}' u_n\, d\tau\right) e^{-iE_n t/\hbar} = i\hbar \sum_n \dot{a}_n\left(\int u_m^* u_n\, d\tau\right) e^{-iE_n t/\hbar}$$

The left-hand integral is the matrix element H'_{mn}. The right-hand one is the orthonormality integral, giving δ_{mn}; this kills all but the mth term in the right-hand summation, whence

$$\dot{a}_m = (i\hbar)^{-1} \sum_n H'_{mn} a_n\, e^{i\omega_{mn}t} \qquad (11.4)$$

where $\omega_{mn} = (E_m - E_n)/\hbar$.

So far everything is exact. The set of equations (11.4) for the coefficients a_m is entirely equivalent to the Schrödinger equation (11.2). We now introduce the perturbation approximation along the same lines as in the stationary case, replacing $\hat{H}'$ by $\lambda\hat{H}'$, and expanding the a_n in powers of λ: $a_n = a_n^{(0)} + \lambda a_n^{(1)} + \cdots$. Equation (11.4) must be true order by order in λ. The zeroth order just gives $\dot{a}_m^{(0)} = 0$, which tells us that all the $a_m^{(0)}$ are constant. This reproduces the stationary case in the absence of $\hat{H}'$. The first-order equation is

$$\dot{a}_m^{(1)} = (i\hbar)^{-1} \sum_n a_n^{(0)} H'_{mn}\, e^{i\omega_{mn}t} \qquad (11.5)$$

which integrates to give

$$a_m^{(1)}(t) = (i\hbar)^{-1} \sum_n a_n^{(0)} \int_0^t H'_{mn}\, e^{i\omega_{mn}t}\, dt' \qquad (11.6)$$

The lower limit on the integral has been chosen to correspond to the time at which the perturbation is switched on, e.g. $t = 0$. For some problems it is better to replace 0 by $-\infty$.

Suppose that prior to the perturbation the system is in some particular stationary state, say $n = k$. Then at $t = 0$, $a_k = 1$ and $a_n = 0$,

$n \neq k$. In that case, remembering that all the a_n^0 are constant, equation (11.6) reduces to

$$a_m(t) = (i\hbar)^{-1} \int_0^t H'_{mk} e^{i\omega_{mk}t'} \, dt' \qquad (11.7)$$

where we have dropped the superscript (1) for convenience. Comparing equation (11.7) with equation (11.4) reveals the physical significance of the perturbation approximation. It amounts to treating all the $a(t)$ in the summation as remaining constant at their initial values, which is a good approximation if the disturbance causes negligible change.

The quantity $|a_m(t)|^2$ given by equation (11.7) represents the probability of finding the system in state m at time t. As the system started out in state k, this tells us the probability that a *transition* will have occurred by time t from state k to state m. Quantum uncertainty prevents our knowing, or even attaching meaning to, the exact moment of transition.

11.2 HARMONIC PERTURBATION

As a simple yet very important application, suppose the perturbation has the form

$$H'(r, t) = \bar{H}(r) \cos \omega t \qquad (11.8)$$

This might, for example, represent an atom being stimulated by an electromagnetic wave or a periodically varying electric field.

Substituting equation (11.8) into equation (11.7) and performing the integral gives

$$a_m = -\frac{\bar{H}_{mk}}{2\hbar} \left[\frac{e^{i(\omega_{mk} - \omega)t} - 1}{\omega_{mk} - \omega} + \frac{e^{i(\omega_{mk} + \omega)t} - 1}{\omega_{mk} + \omega} \right] \qquad (11.9)$$

Transitions will be appreciable if $\omega \approx \pm \omega_{mk}$ because one of the denominators of the two terms in equation (11.9) then becomes very small and the term itself becomes very large. We may then neglect the other term.

Suppose $\omega \approx \omega_{mk}$. Then

$$|a_m|^2 \approx |\bar{H}_{mk}|^2 \sin^2[\tfrac{1}{2}(\omega_{mk} - \omega)t]/\hbar^2(\omega_{mk} - \omega)^2 \qquad (11.10)$$

The factor $\sin^2[\tfrac{1}{2}(\omega_{mk} - \omega)t]/(\omega_{mk} - \omega)^2$ is plotted as a function of ω in Fig. 11.1, from which it may be seen that function (11.10) is

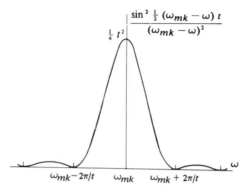

Fig. 11.1 The peak of the function has height $\frac{1}{4}t^2$ and width (at the base) of $4\pi/t$. It therefore sharpens with increasing t. The area beneath the peak is αt.

strongly peaked around $\omega = \omega_{mk}$. This is reminiscent of the phenomenon of resonance, which is indeed what it is, with the quantum matter waves being driven sympathetically by the harmonic perturbation. The width of the profile is proportional to t^{-1} so that if the transition is slow (t large) the range of any effective driving frequency ω is restricted to a very narrow interval around $\omega = \omega_{mk}$. Then

$$\hbar\omega \approx \hbar\omega_{mk} \equiv E_m - E_k \tag{11.11}$$

As $\omega > 0$, $E_m > E_k$, so the system absorbs a quantum of energy from the external disturbing agency and jumps up from the initial energy level E_k to a higher (excited) level E_m. We therefore recognize equation (11.11) as Bohr's frequency condition (equation (1.6)).

If $\omega \approx -\omega_{mk}$ we must reverse the signs of ω_{mk} in equations (11.10) and (11.11). In that case $E_k > E_m$, corresponding to a downward transition from an initial excited state k to a lower state m, with the attendant emission of a quantum $\hbar\omega$ of energy.

11.3 ENERGY–TIME UNCERTAINTY RELATION AND FERMI'S GOLDEN RULE

Equation (11.11) is a statement of the classical law of conservation of energy. The jump in energy of the quantum system, $E_m - E_k$, is paid for by the energy of the quantum, $E = \hbar\omega$. However, the finite width of the function shown in Fig. 11.1 implies that equation (11.11) need only be approximately correct, and transitions are still possible even if $\hbar\omega$ does not exactly equal the energy difference $E_m - E_k$. Thus the law of energy conservation can be slightly violated in quantum mechanics. The width of the profile in Fig. 11.1 shows by how much. The value of the function is appreciable over a range in ω of about $1/t$, or an energy range $\Delta E \sim \hbar/t$. Because there is no way of knowing when during the interval 0 to t the transition takes place, we can regard t as the duration of the transition and write it Δt. We then have

$$\Delta E \, \Delta t \sim \hbar \qquad (11.12)$$

which is known as the energy–time uncertainty relation and should be compared with the momentum–position uncertainty relation (equation (1.11)).

Sometimes equation (11.12) is described by saying that for a short interval Δt an amount of energy ΔE can can be 'borrowed', so long as it is returned at the end of the duration. The shorter the interval, the bigger the loan. Although this crude notion should not be taken too seriously, it can assist in building up a physical image of a number of quantum phenomena. For example the tunnelling of a particle through a thin barrier, described in section 4.5, could alternatively be viewed as a high-jump over the barrier on borrowed energy.

The fact that energy conservation does not have to be rigorously satisfied has two important consequences. First it means that if the perturbation contains a mixture of frequencies in a range around $\omega = \omega_{mk}$, they will all contribute to the *total* transition probability. In that case we have to sum the individual probabilities $|a_m|^2$ over all the frequencies in the perturbation.

Second, because the system is permitted to make a transition to a state that lies anywhere within a range ΔE of energies around the classical permitted value (and the faster it makes it the wider the range) often means that, in practice, the system has available many final states. For example, in the case of an atomic electron being ejected entirely from an atom by the absorption of a photon with energy greater than the ionization energy (the photoelectric effect), the final state of the electron belongs to a whole continuum of energy levels. If we are only interested in the *total* probability for a transition to *any* of the available final states we must once again sum (or integrate) the individual probabilities over all available final states.

Thus instead of computing merely $|a_m|^2$, we are interested in $\sum_m |a_m|^2$. If the states are so closely spaced as to form, or approximate, a continuum (as in the photoelectric effect, for example) we may envisage a *density* of final states, i.e. a number per unit frequency interval ω_{mk}, which we denote $\rho(\omega_{mk})$. We may then make the replacement

$$\sum_m |a_m|^2 \rightarrow \int_{-\infty}^{\infty} |a_m|^2 \rho(\omega_{mk}) \, d\omega_{mk} \qquad (11.13)$$

Now we have seen that $|a_m|^2$ is sharply peaked around the value $\omega_{mk} = \omega$. Generally ρ, and also the matrix element $\bar{H}_{mk}$ which appears in equation (11.10), are much more slowly varying functions of ω_{mk}. In that case it is a good approximation to replace these functions by their values at $\omega_{mk} = \omega$ and to remove them from the integral in equation (11.13). We are then left with

$$\frac{|\bar{H}_{mk}(\omega)|^2}{4\hbar^2} \rho(\omega) \int_{-\infty}^{\infty} \frac{\sin^2 \frac{1}{2}(\omega_{mk} - \omega)t}{[(\omega_{mk} - \omega)/2]^2} \, d\omega_{mk}$$
$$= \pi |\bar{H}_{mk}|^2 \rho t / 2\hbar^2 \qquad (11.14)$$

for the total transition probability. The integral we have computed here is proportional to the area under the curve shown in Fig. 11.1. A useful quantity is the *transition rate*, w, defined to be the transition probability per unit time. From equation (11.14) this is

$$w = \pi |\bar{H}_{mk}|^2 \rho(\omega) / 2\hbar^2 \qquad (11.15)$$

a statement known as Fermi's golden rule, which has a wide range of applications.

11.4 EMISSION AND ABSORPTION OF ELECTROMAGNETIC RADIATION BY ATOMS

A proper treatment of this topic requires the quantization of the electromagnetic field, but the correct results can still be obtained by regarding the field as a classical perturbation.

An electromagnetic wave contains an oscillating electric field E. In most practical cases the wavelength is much greater than the size of the atom, so we may ignore the small variation in E across the atom and write $E = E_0 \cos \omega t$ where E_0 is approximately constant. The interaction energy of this field with a particle of charge $-e$, e.g. an electron, is $-eE \cdot r$. (For slow-moving particles, magnetic effects may be ignored.) The interaction is therefore a harmonic perturbation and we may use the results of section 11.2 with $\bar{H} = -eE_0 \cdot r$. The transition probability for the atom to jump from state k to state m

under this perturbation is given by equation (11.10):

$$e^2 E_0^2 |\langle m|r|k\rangle|^2 \sin^2[\tfrac{1}{2}(\omega_{mk} - \omega)t]/3\hbar^2(\omega_{mk} - \omega)^2 \quad (11.16)$$

where the matrix element H_{mk} has been written in Dirac notation and the factor $\tfrac{1}{3}$ arises because an average has been taken over all orientations of the vector E_0 to correspond to unpolarized radiation.

Suppose the atom is stimulated by a steady flux of radiation. From electromagnetic theory we know that the electric energy density in the wave is $\tfrac{1}{2}\varepsilon_0 E^2$, which averages over a cycle to $\tfrac{1}{4}\varepsilon_0 E_0^2$. The magnetic energy density is the same. Travelling at speed c this represents a total energy flux of $\tfrac{1}{2}\varepsilon_0 E_0^2 c$.

In most practical applications the wave would consist of a range of frequencies superimposed with random phases. We assume the range $\Delta\omega$ is much broader than the sharp peak in Fig. 11.1. The *total* transition probability for all these stimulating frequencies is then given by summing expression (11.16) over $\Delta\omega$. As the frequencies ω form a continuum, this step is accomplished by replacing the total energy flux by the energy flux per unit frequency interval, denoted $I(\omega)$, and integrating over $\Delta\omega$. The quantity $I(\omega)$ is known as the *intensity* of the radiation.

Making these replacements in expression (11.16) the total transition probability is

$$\frac{2e^2}{3\varepsilon_0 \hbar^2 c} \int_{\Delta\omega} \{I(\omega)|\langle m|r|k\rangle|^2 \sin^2[\tfrac{1}{2}(\omega_{mk} - \omega)t]/(\omega_{mk} - \omega)^2\} \, d\omega$$
$$(11.17)$$

Once again we argue that the final factor is sharply peaked around $\omega = \omega_{mk}$ so that, to good approximation, the other more slowly varying functions of ω may be replaced by their values at ω_{mk} and removed from the integral. The integral may then be performed by assuming $\Delta\omega$, being much broader than the peak in the integrand, may be extended to the range $-\infty$ to ∞ with negligible error. The integral yields $\pi t/2$. With all these approximations, then, we get for the total transition *rate*

$$w = \pi e^2 I(\omega_{mk})|\langle m|r|k\rangle|^2/3\varepsilon_0 \hbar^2 c \quad (11.18)$$

This result applies both to upward transitions (stimulated absorption of radiation) if $E_m > E_k$, or downward transitions (stimulated emission of radiation) if $E_m < E_k$. Classically, er is known as the electric dipole moment, so equation (11.18) is called the rate for electric dipole transitions. It may happen that the dipole matrix element vanishes for some choice of initial and final states. In this case the approximations used above break down. We may then have to take into account magnetic processes or the spatial variation of the wave over the atom. This leads to matrix elements analogous to magnetic dipoles, electric quadrupoles, etc. The transition rates for these higher-order effects are usually much lower. In addition, terms from higher orders of perturbation theory may be important. Such terms may be interpreted as the simultaneous emission or absorption of more than one photon at a time.

A good example concerns spherically symmetric systems. Because r is an odd function the matrix element $\langle m|r|k \rangle$ vanishes unless states $|m\rangle$ and $|k\rangle$ have opposite parity. This leads to the *selection rule* that for electric dipole radiation the quantum number l must change by one: $\Delta l = \pm 1$. Similarly one may show that $\Delta m = 0$ or ± 1 (see Exercise 11.4). These results may be used to deduce the fact that the photon carries away angular momentum of one unit ($\hbar$) and is therefore a 'spin-one' particle.

11.5 SPONTANEOUS EMISSION OF RADIATION

The stimulated emission and absorption of radiation both have close classical analogues. If a simple harmonic oscillator is driven by a sinusoidal electromagnetic field it can gain or lose energy depending on phase angles. However, we also know from classical theory that an oscillating electric charge radiates energy spontaneously without the need for a driving force. Classical electrodynamics describes the power radiated in all directions by an accelerating charged particle in terms of the Larmor formula

$$P = e^2 \ddot{r}^2 / 6\pi\varepsilon_0 c^3 \tag{11.19}$$

where $\ddot{r}$ is the acceleration. For an oscillating charge $r = r_0 \cos \omega t$. If we average P over a cycle we get

$$e^2 \omega^4 r_0^2 / 12\pi\varepsilon_0 c^3 \tag{11.20}$$

The quantum formula follows by replacing r_0^2 by $|2\langle m|r|k\rangle|^2$. (The factor of 2 arises because $\cos \omega t$ represents a stimulation containing two Fourier frequency components, $\frac{1}{2}\exp(i\omega t)$ and $\frac{1}{2}\exp(-i\omega t)$, and quantum mechanically only the former contributes to the emission process.)

Dividing by the energy of a typical photon, $\hbar\omega_{mk}$, expression (11.20) yields for the spontaneous transition rate

$$w_s = e^2 \omega_{mk}^3 |\langle m|r|k\rangle|^2 / 3\pi\varepsilon_0 \hbar c^3 \qquad (11.21)$$

The reason that formula (11.21) did not emerge from the quantum treatment of section 11.4 is because the electromagnetic field was treated classically there. If the field is quantized spontaneous emission can be described without recourse to classical electrodynamics.

The existence of spontaneous emission is important for thermodynamics and statistical mechanics. According to the Boltzmann law, if there is a large collection of identical systems in thermodynamic equilibrium at temperature T, then

$$n(k)/n(m) = e^{-E_k/kT} / e^{-E_m/kT} \qquad (11.22)$$

where $n(k)$, $n(m)$ are the average number of systems in states k and m respectively. Without spontaneous emission (and ignoring collisional de-excitation) we would have $n(k) = n(m)$. The effect of the additional downward transition rate is to keep a majority of systems in the lower energy state, as required by equation (11.22). Using the fact that in equilibrium the upward and downward transition rates must exactly balance, we can write

(rate of stimulated emission + rate of spontaneous emission) × $n(k)$
= (rate of stimulated absorption) × $n(m)$

where $E_k > E_m$. We may substitute the Boltzmann relation (equation 11.22) for $n(k)$ and $n(m)$ and use formulae (11.18) and (11.21) for the transition rates. The matrix elements cancel to leave, finally

$$I(\omega) = \frac{\hbar\omega^3}{\pi c^2(e^{\hbar\omega/kT} - 1)} \qquad (11.23)$$

where $\omega = (E_k - E_m)/\hbar$. This is the famous Planck spectrum for thermal equilibrium radiation. The argument just given could be inverted, using Planck's already known formula to relate the transition rates of spontaneous and stimulated emission (compare equation (11.18) with equation (11.21). This was the procedure used by Einstein in 1917 prior to the development of quantum mechanics. He introduced the so-called A and B coefficients as a measure of the emission and absorption rates. These coefficients are essentially given by equations (11.18) and (11.21).

The existence of spontaneous emission implies that all excited states of electrical systems are unstable and will eventually decay to the ground state by photon emission. Quantum systems thus comply with the fundamental principle that physical systems tend to seek out their state of lowest energy and maximum entropy. The lifetime τ of an excited state is $\approx w_s^{-1}$. From the energy–time uncertainty relation (equation (11.12)) this implies that the emitted photon has an energy uncertainty of $\Delta E \approx \hbar w_s$. This shows up in the spectra of spontaneously radiating systems as a finite width of all spectral lines. The frequency spread is $\approx \tau^{-1}$ so the longer the lifetime the sharper the line.

These ideas find a close parallel with classical electrodynamics. A simple harmonic oscillator decaying by radiative emission damps in amplitude like $\exp(-t/\tau)$ where $\tau \approx$ half-life for decay. A Fourier analysis of damped simple harmonic motion $\exp(-t/\tau)\cos\omega_0 t$ gives the frequency spectrum $1/[(\omega - \omega_0)^2 + (1/\tau)^2]$, representing the emission of frequencies centred around ω_0 with a spread $\approx \tau^{-1}$.

APPENDIX A

The one-dimensional infinite array of delta wells (Kronig–Penney model)

In Chapter 3 we discussed the single and triple delta wells, and this appendix takes the treatment on to higher numbers of wells. The point of this is to model the behaviour of electrons in crystalline lattices. These lattices are of course three dimensional, and the potential in the neighbourhood of an atom is not very much like a delta function; nevertheless we shall give details only of a one-dimensional array of delta functions, providing references for the ambitious reader who wishes to pursue the matter further.

Extension of the theory from $N = 3$ (the triple well) to an infinite array $(N \to \infty)$ produces the interesting result that for a given value of s, defined in equation (3.24), the allowed states (both bound and unbound) are grouped in bands of allowed energies. Each band encompasses N allowed states. When we dealt with $N = 3$, this banding was not apparent; all that Fig. 3.5 shows is a group of three allowed states. Also, in Chapter 3 we restricted attention to bound states, and we shall take up the account at that point. In the case of $N = 3$, the forms of the three eigenstates are shown in Fig. 3.4. The eigenstates for $N = 4$, calculated in just the same way, are shown in Fig. A.1.

Figure A.1 suggests a periodicity in each of the four wave functions for $N = 4$; if one pursues this to higher values of N, the periodicity becomes a marked feature, and it is best to use a notation which allows this to flow naturally from the algebra. What we want is a more general form for equations (3.26), and we proceed by seeking

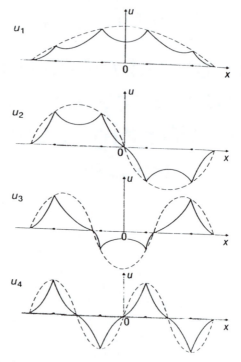

Fig. A.1 This shows the four allowed wave functions for the bound states of a quadruple delta well, with the periodicity indicated by the dashed curves.

an appropriate solution of equation (3.18):

$$\frac{\mathrm{d}^2 u}{\mathrm{d} x^2} = \beta^2 u \tag{A.1}$$

We could start by trying equation (3.19) as for the triple well:

$$u_1 = A \exp(\beta x) + B \exp(-\beta x) \tag{A.2}$$

However, in view of the apparent periodicity remarked above, we prefer to use a more flexible way of writing the solution to equation (A.1), appropriate as $N \to \infty$:

$$u_n = \exp(\mathrm{i} k n b)\{A \exp[\beta(x - nb)] + B \exp[-\beta(x - nb)]\} \tag{A.3}$$

for the region $nb \leqslant x \leqslant (n + 1)b$. (There is a mathematical theorem, the Bloch theorem, which validates this form but we shall not prove it here.) In this equation (A.3) n labels the regions between wells. The constants k, A, and B are independent of n and can be found by applying the boundary conditions. These are essentially the same as equations (3.21) and (3.22) except that they apply at all the N wells:

$$u_{n+1}(x) = u_n(x) \quad \text{at} \quad x = (n+1)b \quad \text{(A.4)}$$

and

$$\frac{du_{n+1}}{dx} - \frac{du_n}{dx} = -\frac{2s}{b}u_n \quad \text{at} \quad x = (n+1)b \quad \text{(A.5)}$$

(We are using s as defined in equation (3.24) in order to simplify the notation.) It is now straightforward algebra to apply these boundary conditions to the solution (A.3) to find the pair of equations

$$A\exp(\beta b) + B\exp(-\beta b) = (A+B)\exp(ikb) \quad \text{(A.6)}$$

and

$$\beta A\exp(\beta b) - \beta B\exp(-\beta b) - \beta A\exp(ikb) + \beta B\exp(ikb)$$
$$= (2s/b)(A+B)\exp(ikb) \quad \text{(A.7)}$$

The constants A and B may now be eliminated from equations (A.6) and (A.7), leaving an equation for k which is most conveniently written in terms of the dimensionless parameters s and ε defined in Chapter 3:

$$\cos(kb) = \cosh[(-\varepsilon)^{1/2}s] - \frac{1}{(-\varepsilon)^{1/2}}\sinh[(-\varepsilon)^{1/2}s] \quad \text{(A.8)}$$

Now the cosine function on the left can only take values within the range $+1$ to -1, and that range can be covered by allowing all values of kb in the range 0 to π. The function $\varepsilon(kb)$, obtained numerically from equation (A.8) in that range, for $s = 2$, is shown in the left-hand segment of Fig. A.2. It invites comparison with Fig. 3.5 for the triple well, from which one can see that the bound states for $s = 2$ for the triple well all lie within the boundaries of the lowest band for the infinite array.

The limiting values then allow the lower and upper edges of the allowed band of energies to be calculated. The lower and upper edges correspond to equation (A.8) with the left hand side set to ± 1; the two resulting equations can then be rearranged for easy computation as

$$s = y\tanh(y/2) \quad \text{(lower edge for } s \geqslant 0) \quad \text{(A.9)}$$

and

$$s = y\coth(y/2) \quad \text{(upper edge for } s \geqslant 2) \quad \text{(A.10)}$$

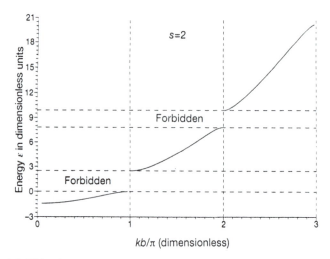

Fig. A.2 This shows how ε depends on kb for the choice $s = 2$ for the dimensionless separation parameter. It is plotted numerically from equation (A.8) for the range $kb = 0$ to $kb = \pi$ and the limiting points can be fairly readily seen to be $\varepsilon = -1.44$ for $kb = 0$ and $\varepsilon = 0$ for $kb = \pi$. (The fact that ε approaches zero at the upper end of the range is an accidental consequence of the choice $s = 2$; if we had chosen $s = 3$, the upper limit of ε would have been -0.74.) The two higher-energy segments are calculated from equation (A.12) for the ranges $kb = \pi$ to $kb = 2\pi$ and $kb = 2\pi$ to $kb = 3\pi$.

where

$$y \equiv (-\varepsilon)^{1/2} s \tag{A.11}$$

We have so far only considered the bound states, i.e. those for which ε is negative. It is no more difficult to follow a similar mathematical procedure for unbound states, i.e. those for which ε is positive. This is necessary in order to compute the upper edge for values of s less than 2, but it is also interesting because it reveals that there is a whole hierarchy of higher-energy allowed bands above the one we have described; the physical origin of this unexpected result will be discussed below. Meanwhile, the reader may care to confirm that, for positive values of ε, the equation corresponding to equation (A.8) is

$$\cos(kb) = \cos[(+\varepsilon)^{1/2} s] - \frac{1}{(+\varepsilon)^{1/2}} \sin[(+\varepsilon)^{1/2} s] \tag{A.12}$$

Equations for the upper and lower band edges can be obtained from this equation in much the same way as equations (A.1) to (A.11) were obtained from equation (A.8) by setting the left-hand side to ± 1.

Figure A.2 shows the first three allowed bands of energy for the chosen value of $s = 2$. Now something has to be asked regarding the forbidden energy gaps between the allowed bands. The mathematics tells us that energies in that range are forbidden, and that the integral values of kb/π are associated with those gaps. However, what is the physics of this? The answer is to look at the forms of the probability distributions calculated from the wave functions corresponding to values of kb/π just above and just below an integral value. The wave functions are both found to be standing waves and their shapes can be laboriously deduced from equation (A.3) together with the calculated value of kb and equation (A.6) or (A.7) which can be used to eliminate B. Figure A.3 shows the shapes corresponding to the bottoms and tops of the second and third allowed bands shown in Fig. A.2. The physics then is that band edges correspond to *standing*

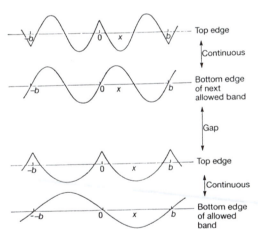

Fig. A.3 The forms of the wave functions are standing waves at the tops and bottoms of allowed bands. The figure illustrates the second and third band edges and shows how the standing waves pile up probability either at a well or away from a well. There is an energy difference because in one case there is potential energy while in the other there is not. At the bottom of the allowed band (top of a forbidden gap) the wave function is zero at the well positions.

waves which pile up particles *either* at the wells *or* between the wells, and there is clearly an energy difference between these two possibilities. Symmetry arguments ensure that the distribution is symmetrical with respect to the array; there may be peaks *at* wells or halfway between wells, but not at arbitrary locations. This means that for a given value of k, there is an energy discontinuity (a band gap) because an energy between the energies of the allowed standing waves would not correspond to an acceptable wave function. Within an allowed band, the form of the wave function is not a standing wave but a travelling wave.

More mathematical details of the model are well described in *Wave Mechanics of Crystalline Solids* by R. A. Smith, second edition published in 1969 by Chapman & Hall. It is of course desirable to extend these considerations into three dimensions and to use more realistic well characteristics. This is beyond the remit of the present volume, but the interested reader might care to refer to an article by J. C. Slater entitled 'A soluble problem in energy bands', published in 1952 in the *Physical Review*, volume 87, pages 807–835.

APPENDIX B

Spherical harmonics

The functions $Y_{lm}(\theta, \phi)$ introduced in Chapter 7 arise in a wide range of physical problems (not just quantum mechanics) and their properties have been much studied.

We first introduce the Legendre polynomials. Legendre's equation is

$$\frac{d}{dx}\left[(1 - x^2)\frac{dP(x)}{dx}\right] + \lambda P(x) = 0 \tag{B.1}$$

(λ constant). Equation (B.1) may be solved by the method of series. Substituting

$$P(x) = x^\alpha \sum_{k=0}^{\infty} a_k x^k \tag{B.2}$$

we get

$$\sum_{k=0}^{\infty} a_k\{(k + \alpha)(k + \alpha - 1)x^{k + \alpha - 2} + [\lambda - (k + \alpha)(k + \alpha + 1)]x^{k + \alpha}\} = 0 \tag{B.3}$$

This can only be true for all x if the coefficient of each power of x vanishes. The lowest power is $x^{\alpha - 2}$. It has coefficient $a_0\alpha(\alpha - 1)$. Since $a_0 \neq 0$ by assumption, we conclude $\alpha = 0$ or 1. The coefficient of a general power $x^{k + \alpha}$ is

$$a_{k+2}(k + 2 + \alpha)(k + 1 + \alpha) + a_k[\lambda - (k + \alpha)(k + 1 + \alpha)]$$

which vanishes if

$$\begin{aligned} a_{k+2}/a_k &= [k(k + 1) - \lambda]/(k + 1)(k + 2) &&(\alpha = 0) \\ &= [(k + 1)(k + 2) - \lambda]/(k + 2)(k + 3) &&(\alpha = 1) \end{aligned} \tag{B.4}$$

Thus each coefficient is related recursively to the lower order coefficients. Because of the jump of 2 in the value of k there will be two series, the $\alpha = 0$ series containing only even powers of x (and hence of even parity) and the $\alpha = 1$ series containing only odd powers of x (and of odd parity).

For large k the ratio (B.4) approaches 1. This is the same asymptotic behaviour as the function $(1 - x^2)^{-1}$. Such a function diverges at $x = \pm 1$. Since a divergent wave function is unacceptable on normalization grounds, a solution of the form (B.2) can only be entertained if the power series is prevented from reaching $k = \infty$. This may be accomplished by restricting the value of λ to the set

$$\lambda = l(l + 1) \begin{cases} l = 0, 1, 2, 3, \ldots & (\alpha = 0) \\ l = 1, 2, 3, \ldots & (\alpha = 1) \end{cases}$$

so that at the term in the series for which $k = l$ ($\alpha = 0$) or $k = l - 1$ ($\alpha = 1$) the ratio (B.4) will vanish. As each coefficient in the series is proportional to the previous even (or odd) order coefficient, if one coefficient vanishes, so will all the remaining coefficients in the series. The resulting polynomial solutions, well behaved for all x, are known as Legendre polynomials and denoted $P_l(x)$, $l = 0, 1, 2, 3, \ldots$. The first few are

$$\left. \begin{array}{l} P_0(x) = 1 \\ P_1(x) = x \\ P_2(x) = \frac{1}{2}(3x^2 - 1) \\ P_3(x) = \frac{1}{2}(5x^3 - 3x) \end{array} \right\} \tag{B.5}$$

where the overall factor has been chosen to make $P_l(1) = 1$ in each case. Note that the parity of these polynomials is $(-1)^l$.

Legendre polynomials may also be generated using the formulae

$$P_l(x) = \frac{1}{2^l l!} \frac{d^l}{dx^l} (x^2 - 1)^l \tag{B.6}$$

or

$$\sum_{l=0}^{\infty} r^{-l-1} P_l(\cos \theta) = (r^2 - 2r \cos \theta + 1)^{-1/2} \tag{B.7}$$

Closely related to P_l are the associated Legendre polynomials which depend on two indices, l and m. Denoted $P_l^m(x)$, they may be generated by the operation

$$P_l^m(x) = (1 - x^2)^{m/2} \frac{d^m P_l}{dx^m} \quad m \geqslant 0 \tag{B.8}$$

For $m < 0$

$$P_l^m = (-1)^m [(l+m)!/(l-m)!] P_l^{-m}$$

It follows from the fact that P_l is an lth degree polynomial that $-l \leqslant m \leqslant l$. It is easy to show that if P_l satisfies Legendre's equation (B.1) then P_l^m satisfies equation (7.20) (with $\hbar = 1$).

The functions P_l^m with fixed m form an orthogonal set on the interval $-1 \leqslant x \leqslant 1$

$$\int_{-1}^1 P_{l'}^m(x) P_l^m(x)\, dx = \frac{2}{2l+1} \frac{(l+m)!}{(l-m)!} \delta_{ll'} \tag{B.9}$$

Normalization of P_l^m thus requires multiplication by the factor $[(l+\frac{1}{2})(l-m)!/(l+m)!]^{1/2}$. They satisfy the recurrence relation

$$(l+m)P_{l-1}^m + (l-m+1)P_{l+1}^m - (2l+1)xP_l^m = 0 \tag{B.10}$$

Spherical harmonics are functions of two variables and are defined by equation (7.22). They are orthonormal in the following sense

$$\int_0^{2\pi} d\phi \int_0^{\pi} \sin\theta\, d\theta\, Y_{l'm'}^*(\theta, \phi) Y_{lm}(\theta, \phi) = \delta_{ll'}\delta_{mm'} \tag{B.11}$$

Several examples are given in Table 7.1 in section 7.2
The following property is useful

$$\sum_{m=-l}^l Y_{lm}^*(\theta, \phi) Y_{lm}(\theta', \phi') = \frac{2l+1}{4\pi} P_l(\cos\alpha) \tag{B.12}$$

where α is the direction between the angles (θ, ϕ) and (θ', ϕ').

APPENDIX C

The Dirac delta function

This mathematical object, denoted $\delta(x - x')$, was introduced in section 6.3. It has the following properties

$$\delta(x - x') = 0 \qquad x \neq x'$$
$$= \infty \qquad x = x'$$

$$\int_a^b \delta(x - x')\,dx = 1 \qquad\qquad\qquad \text{(C.1)}$$

so long as $a < x' < b$. For any smooth function f

$$\int_a^b f(x)\delta(x - x')\,dx = f(x') \qquad\qquad\qquad \text{(C.2)}$$

so δ 'projects out' the value of f at $x = x'$.

Moreover, the identities

$$\delta(x) = \delta(-x) \qquad\qquad\qquad \text{(C.3)}$$
$$x\,d\delta/dx = -\delta(x) \qquad\qquad\qquad \text{(C.4)}$$
$$x\delta(x) = 0 \qquad\qquad\qquad \text{(C.5)}$$

can all be proved by multiplying both sides by $f(x)$ and integrating.

The δ function may be considered as the derivative of the Heaviside step function

$$\left.\begin{aligned}\theta(x - x') &= 0 \quad x < x'\\ &= 1 \quad x > x'\end{aligned}\right\} \qquad\qquad \text{(C.6)}$$

To prove this, put $d\theta/dx$ in place of δ in (C.2) and integrate by parts:

$$\int_a^b \frac{d\theta(x-x')}{dx} f(x)\,dx = f(b) - \int_a^b \theta(x-x')\frac{df(x)}{dx}\,dx$$

$$= f(b) - \int_{x'}^b df(x) = f(x')$$

A useful representation of $\delta(x)$ is as follows:

$$\delta(x) = \frac{1}{2\pi}\int_{-\infty}^\infty e^{ikx}\,dk$$

In three dimensions one may define

$$\delta^3(\mathbf{r}) = \delta(x)\delta(y)\delta(z) = \frac{1}{8\pi^3}\int_{-\infty}^\infty\int_{-\infty}^\infty\int_{-\infty}^\infty e^{i\mathbf{k}\cdot\mathbf{r}}\,dx\,dy\,dz$$

Suppose a collection of function $\phi_n(x)$ form a complete ortho-normal set. We may then expand an arbitrary function

$$\psi = \sum_n c_n\phi_n \tag{C.7}$$

where $c_n = \langle\phi_n|\psi\rangle \equiv \int\phi_n^*\psi\,dx$. Substituting this back into (C.7) yields

$$\psi(x) = \sum_n \phi_n(x)\int\phi_n^*(x')\psi(x')\,dx'$$

$$= \int\left[\sum_n \phi_n(x)\phi_n^*(x')\right]\psi(x')\,dx' \tag{C.8}$$

Comparing (C.8) with the definition (C.2) of the delta function, we deduce

$$\sum_n \phi_n(x)\phi_n^*(x') = \delta(x-x') \tag{C.9}$$

This is known as the closure relation.

APPENDIX D

Useful quantities

Planck's constant
$$h = 6.63 \times 10^{-34} \, \text{J s}$$
$$\hbar \equiv h/2\pi = 1.05 \times 10^{-34} \, \text{J s}$$

Electronic charge
$$e = 1.60 \times 10^{-19} \, \text{C}$$

Electric constant
$$\varepsilon_0 = 8.85 \times 10^{-12} \, \text{F m}^{-1}$$

Rest mass of electron
$$m_e = 9.11 \times 10^{-31} \, \text{kg}$$

Rest mass of proton
$$m_p = 1.67 \times 10^{-27} \, \text{kg}$$

Speed of light
$$c = 3.00 \times 10^8 \, \text{m s}^{-1}$$

Electron volt
$$1 \, \text{eV} = 1.60 \times 10^{-19} \, \text{J}$$
$$(1 \, \text{MeV} = 10^6 \, \text{eV})$$

Bohr radius
$$a_0 \equiv 4\pi\varepsilon_0\hbar^2/me^2 = 5.29 \times 10^{-11} \, \text{m}$$

Bohr magneton
$$\mu_B \equiv e\hbar/2m_e = 9.27 \times 10^{-24} \, \text{J T}^{-1}$$

Fine structure constant
$$\alpha \equiv e^2/4\pi\varepsilon_0\hbar c = 1/137.036$$

APPENDIX E

Useful integrals

$$\int x^n e^{-x/a}\,dx = -e^{-x/a}\left[ax^n + \sum_{k=1}^{n} n(n-1)\cdots(n-k+1)a^{k+1}x^{n-k}\right]$$

$$\int_0^\infty x^n e^{-x/a}\,dx = n!a^{n+1}$$

$$\int_{-\infty}^\infty x^{2n}e^{-a^2x^2}\,dx = \begin{cases} \pi^{1/2}/a & n=0 \\[2mm] \dfrac{1\times 3\times 5\cdots(2n-1)\pi^{1/2}}{2^n a^{2n+1}} & n=1,2,3,\ldots \end{cases}$$

$$\int_{-\infty}^\infty e^{-a^2x^2\pm bx}\,dx = (\pi^{1/2}/a)e^{b^2/4a^2}$$

$$\int_1^\infty e^{-x^2}\,dx = 0.157\pi^{1/2}$$

$$\int_{-\infty}^\infty e^{ikx-\beta|x|}\cos\alpha x\,dx = 2\beta(\alpha^2+\beta^2+k^2)/[(\alpha^2+\beta^2+k^2)^2-4\alpha^2k^2]$$

$$\int_{-a}^a \sin^2(n\pi x/2a)\,dx = \int_{-a}^a \cos^2(n\pi x/2a)\,dx = a$$

$$\int_{-a}^a x^2\sin^2(n\pi x/2a)\,dx = \tfrac{1}{3}a^3[1-6(-1)^n/n^2\pi^2]$$

$$\int_{-a}^a x^2\cos^2(n\pi x/2a)\,dx = \tfrac{1}{3}a^3[1+6(-1)^n/n^2\pi^2]$$

$$\int_{-\infty}^\infty x^{-2}\sin^2 x\,dx = \pi$$

APPENDIX F

Exercises

A list of relevant integrals is provided in Appendix E.

Chapter 1

1 A photon strikes a stationary electron and scatters exactly backwards along the line of its incoming path. Treating the system as in Newtonian mechanics, apply the conservation of energy and momentum to show that the increase in wavelength of the photon is $2h/m_e c$, where m_e is the rest mass of the electron. [Hint: use equation (1.2) for the energy.]

2 A radio transmitter radiates at a power of 10 kW at a wavelength of 100 m. How many photons does it emit per second?

3 Show that if a particle of mass m is confined in a box of size comparable to h/mc (its so-called Compton wavelength) then the momentum uncertainty is so large that the particle is likely to be moving relativistically.

4 A snooker ball of mass 0.1 kg rests on top of an identical ball and is stabilized by a dent 10^{-4} m wide on the surface of the lower ball. Use the uncertainty principle to estimate how long the system will take to topple, neglecting all but quantum disturbances.

5 Consider the wave packet $\cos(\alpha x)\exp(-\beta|x|)$, where α and β are real positive constants and $\beta \ll \alpha$. Take the Fourier transform of this expression and show that the frequency components are spread over a range $\Delta k \approx \beta$. Hence deduce the uncertainty relation $\Delta x\,\Delta p \approx \hbar$.

6 Consider the wave function $A\exp[i(kr - \omega t)]/r$. Show that the outgoing radial flux of probability is $|A|^2 \hbar k/mr^2$. Compare this with a radial flux of classical particles of speed $v = p/m \equiv \hbar k/m$.

7 Given a one-dimensional wave function

$$\psi(x, t) = a^{-1/2} \cos(\pi x/2a) \exp(-i\pi^2 \hbar t/8ma^2)$$

in the range $-a \leqslant x \leqslant +a$ (and zero outside that range), use equation (1.19) to show that $\langle x \rangle = 0$.
[Note that it can be done by integration if you prefer, but the symmetry of the wave function about the point $x = 0$ is sufficient to deliver the result.]

Go further to find $\langle x^2 \rangle$.
[Note that only integration will do in this case, but the integral you need is given in Appendix E. The answer is $(1 - 6/\pi^2)a^2/3 = 0.131a^2$.]

Chapter 2
1 Using the stationary state wave functions (equation (2.12)) for the infinite square-well problem, calculate $\langle x^2 \rangle$.
2 A particle moves in the potential

$$\begin{aligned}
V &= 0 & |x| < a \\
&= V_0 & a < |x| < b \\
&= \infty & |x| > b
\end{aligned}$$

with energy $E > V_0$. Solve for the stationary state wave functions and find an equation for the energy levels. Show that as $b \to a$ the levels reduce to the result (2.11). Show also that for $E \gg V_0$ the levels approach (2.11) with a replaced by b.
3 Consider the following '1-D analogue' of the hydrogen atom. A particle of mass m and charge e is constrained to move along the x axis with a potential energy

$$\begin{aligned}
V(x) &= \infty & \text{for } x < 0 \\
&= -e^2/4\pi\varepsilon_0 x & \text{for } x \geqslant 0
\end{aligned}$$

(i) Verify that the trial wave function $Ax \exp(-bx)$ is indeed a satisfactory solution of the Schrödinger equation provided uniquely that b is suitably chosen. Show that the unique choice of b is $b = me^2/4\pi\varepsilon_0 \hbar^2$.
(ii) Prove further that the corresponding bound state energy is $E = -me^4/2(4\pi\varepsilon_0 \hbar)^2$.
(iii) Finally, prove that the normalization constant A is given by $A = 2b^{3/2}$.
[You can use integration by parts to perform the integral.]

4 Examine the solutions of the square-well problem treated in section 2.4. Show that for the odd parity stationary states there is no bound state if $V_0 a^2 < \pi^2 \hbar^2/8m$.

5 A particle of mass m is trapped by the potential

$$
\begin{aligned}
V &= 0 & 0 < x < a \\
&= V_0 & x > a \\
&= \infty & x < 0
\end{aligned}
$$

with energy $E < V_0$. Show that the probability densities of finding the particle at $x = 2a$ and $x = a/2$ are proportional to $\sin^2 \alpha a \exp(-2\beta a)$ and $\sin^2 \alpha a/2$ respectively and therefore that the probability of finding the particle at $x = 2a$ relative to that of finding it at $x = a/2$ is equal to $4\cos^2(\alpha a/2)\exp(-2\beta a)$, where α and β are defined as in section 2.4.

Chapter 3

1 Consider a triplet of *finite* square wells of depth V_0 and width a, whose centres are separated by a distance $b > a$. Write down the boundary conditions, solve the Schrödinger equation, and sketch the form of the allowed bound states, commenting on the number and energies of such bound states to be expected for a given set of parameters.

2 Verify equations (3.26a) and (3.26b). You may be surprised at the labour involved in this!

Chapter 4

1 What is the probability of a particle which moves in a simple harmonic oscillator potential being found outside the region accessible to it classically?

2 Consider a charged particle (with mass m and charge q) undergoing one-dimensional simple harmonic motion described quantum mechanically, initially in zero field.

(i) How is the potential $V(x) = \frac{1}{2}kx^2$ changed when an electric field E is applied in the positive x direction?

(ii) Make the substitution $y = x - qE/k$ in the Hamiltonian operator and show that it reduces to the form

$$
\hat{H} = -\frac{\hbar^2}{2m}\frac{\partial^2}{\partial y^2} + \frac{1}{2}ky^2 - q^2E^2/2k
$$

(iii) Hence deduce that the spacing of energy levels is unchanged by the application of the field.
(iv) Finally deduce by finding the minimum in the potential energy that the position of the centre of the probability distribution has shifted by qE/k in the positive x direction.

3 Use parity arguments to show that the energy levels for the motion of a particle in the potential

$$V(x) = \infty \qquad x < 0$$
$$= \tfrac{1}{2}Kx^2 \qquad x > 0$$

are given by $(2n + \tfrac{3}{2})\hbar\omega$, $n = 0, 1, 2, \ldots$. Why is the ground state energy in this system higher than for the simple harmonic oscillator?

4 The Hermite polynomials $H_n(z)$ may be obtained from the generating functional

$$e^{-s^2 + 2sz} = \sum_{n=0}^{\infty} H_n(z)s^n/n!$$

by equating equal powers of s. Expand the left-hand side to order s^5 to obtain explicit expressions for $H_n(z)$ up to $n = 5$. Check your answer using equation (4.13).

5 By differentiating the generating functional in Exercise 4.4 with respect to z and s, derive the following relations

$$\frac{dH_n}{dz} = 2nH_{n-1}$$

$$H_{n+1} = 2zH_n - 2nH_{n-1}$$

Hence show that $H_n(z)$ satisfies Hermite's equation

$$\frac{d^2H_n}{dz^2} - 2z\frac{dH}{dz} + 2nH_n = 0$$

6 By squaring the generating functional in Exercise 4.4 and then integrating over all z, derive the normalization integral used in equation (4.14):

$$\int_{-\infty}^{\infty} H_n^2(z)e^{-z^2} dz = \pi^{1/2}2^n n!$$

7 By following a similar procedure to Exercise 4.6. evaluate

$$\int_{-\infty}^{\infty} z^2 H_n^2(z) e^{-z^2} \, dz$$

Hence deduce $\langle x^2 \rangle$ for the harmonic oscillator in state n. Noting that $V = \frac{1}{2} K x^2$ and that $E_n = \langle T \rangle + \langle V \rangle$, where T is the kinetic energy, deduce that $\langle T \rangle = \langle V \rangle$.

8 Show that successive applications of the step-up operator a^+ yields the harmonic oscillator wave functions $u_n(x)$ given by equation (4.14). [Hint: relate the operator $(a^+)^n$ to equation (4.13) by induction.]

9 A hydrogen atom is in its ground state. Calculate the probability that the electron will be found within 10^{-14} m of the nucleus (assumed point like). What is the probability that the electron will be found outside two Bohr radii of the nucleus?

10 Consider a one-dimensional wave packet which is a superposition of waves $\exp(ikx)$ with amplitudes $\exp(-k^2/4\alpha^2)$, where α is a real constant, over the range $-\infty < k < \infty$. Show that this packet describes a particle with momentum uncertainty $\Delta p \approx \hbar \alpha$.

Perform the k integration to obtain $\psi(x, t)$. Compute the appropriate normalization factor, and then examine $P(x, t)$ to show that the packet corresponds to a Gaussian function centred at $x = 0$ with width $\Delta x \approx \alpha^{-1}$ at time $t = 0$. Hence verify the uncertainty relation $\Delta x \, \Delta p \approx \hbar$.

11 In Exercise 4.10 show that the wave packet spreads with time, the centre remaining at $x = 0$, such that a time $t > 0$ $(\Delta x)^2$ has increased by $\hbar^2 t^2 \alpha^2 / m^2 \approx (\Delta p t / m)^2$. (Note: $\Delta p t / m$ is the distance travelled by a classical particle with momentum Δp in time t.)

12 Repeat Exercise 4.10, using instead the amplitude $\exp[-(k - k_0)^2/4\alpha^2]$, $k_0 = $ real constant, and show that this choice corresponds to a wave packet whose centre of mass moves uniformly with speed $\hbar k_0 / m$.

13 A particle of total energy $9V_0$ is incident from the region $x < 0$ on a potential step in which $V = 0$ for $x < 0$ and $V = 5V_0$ for $x > 0$. Prove that the probability that the particle will be transmitted through the step is $24/25$, i.e. 96%.

14 A steady stream of 5 eV electrons impinges on a square-hill barrier produced by a 10 V potential. If half the electrons penetrate the barrier, how thick is it? How many would penetrate if the barrier were (a) twice as thick, (b) twice as high and (c) a 10 V square well of the original thickness rather than a square hill?

15 Consider the three-dimensional problem of a particle of mass m moving in the harmonic oscillator potential

$$V(r) = \tfrac{1}{2}k_x x^2 + \tfrac{1}{2}k_y y^2 + \tfrac{1}{2}k_z z^2$$

Show that the TISE is separable into three equations leading to the eigenstate being the product of three parts and the energies the sum of three parts. (A similar method of 'separation of variables' was used in section 2.1 to separate the space and time parts of the wave function.)

Show that the three-dimensional harmonic oscillator exhibits degeneracy (i.e. states with different wavefunctions having the same energy) if $k_x = k_y = k_z = k$. Show that the degeneracy of the level with energy $\tfrac{9}{2}\hbar\sqrt{(k/m)}$ is 10.

16 Consider a gas of HCl molecules as a collection of one-dimensional SHOs. Assume that $m_H = 1.66 \times 10^{-27}$ kg, $m_{Cl} = 5.81 \times 10^{-26}$ kg, and that the strength of the HCl bond, defined as the constant k in the potential energy $V = \tfrac{1}{2}kx^2$, is $k = 4.1 \times 10^2$ N m^{-1}.

  (i) Evaluate the classical frequency of vibration of the molecule.
  (ii) If the gas is exposed to electromagnetic radiation, what will be this wavelength of absorbed radiation if transitions can only be induced between adjacent quantum states?
  (iii) If the ratio of the number of atoms in states n and m is given in terms of the temperature T by $N_n/N_m = \exp[-(E_n - E_m)/k_B T]$, estimate (taking $k_B = 1.38 \times 10^{-23}$ J K^{-1}) the temperature below which 99% of the molecules will be in the ground state.

17 Starting from equations (4.44) to (4.46), with $A = 1$ and $\kappa = k$ (corresponding to the energy E of the particles being equal to $V_0/2$) find the constants B, C, D and F, and show that the real parts of the wave functions in the three regions reduce to

  $\cos kx - \tanh ka \sin kx$ to the left of the barrier
  $\cosh[k(x-a)]/\cosh ka$ within the barrier
  $\cos[k(x-a)]/\cosh ka$ to the right of the barrier

These are the functions plotted in Fig. 4.4. Also find and plot the imaginary parts of the wave functions in the three regions.

Chapter 5
1 Prove that $d\sigma = |f(\theta)|^2 d\Omega$ as follows. Take the spherical scattered wave $u = (1/r)f(\theta)\exp(ikr)$ representing the scattered wave, and

carry out the manipulation given in equation (1.18). The operator $\mathbf{V}$ is simply d/dr and $u^* = (1/r)f^*(\theta)\exp(-ikr)$. You should find that the number of particles per unit area per second is

$$\frac{\hbar k}{\mu r^2}|f(\theta)|^2$$

This is the number *per unit area* per second, but to make the connection with $d\sigma$ we want the number into the *solid angle* $d\Omega$ per second. All that is necessary is to multiply by the area $r^2 \, d\Omega$, so the number of particles into the solid angle $d\Omega$ per second is

$$\frac{\hbar k}{\mu}|f(\theta)|^2 \, d\Omega$$

In a similar way take the incident plane wave $\exp(ikz)$ and show (with $\mathbf{V} \to d/dz$) that the incident number of particles per unit area per second, J, is $\hbar k/\mu$. Hence from the definitions given as equation (5.1), prove the required result (5.4).

2 Prove equation (5.33) directly by starting out with the input that the wave function inside a totally hard sphere must be zero and that this determines the boundary condition for the outside wave function at the surface of the sphere.

3 Find the cross-section for a weakly repulsive spherical obstacle, i.e. one for which $V_0 < E$.

Chapter 6

1 Verify that the functions $a^{-1/2}\sin(n\pi x/2a)$ are orthonormal in $-a \leqslant x \leqslant a$ for n even. Show that the expansion coefficients of an odd function $u(x)$ which vanishes at $x = \pm a$ defined in this interval are given by the coefficients of a Fourier sine series.

2 A particle confined to a rigid box $(0 < x < a)$ is described at $t = 0$ by the normalized wave function $u(x) = (30/a)^{1/2}(x/a)(1 - x/a)$. Find the expectation value of the energy. Comparing this with the energy levels of the particle, which state does it appear most likely to be in? Find the probability of obtaining the lowest energy level when an energy measurement is carried out.

3 If the particle in Exercise 6.2 has the wave function $2(2/3a)^{1/2}\sin^2(\pi x/a)$, find the expectation value of the energy, $\langle E \rangle$. What is the probability that the particle will be found in (a) the ground state and (b) the first excited state? Multiply these probabilities by the respective energies of those states and compare with $\langle E \rangle$.

4 Show that the eigenvalues of the parity operator $\hat{P}$ are ± 1.

5 A one-dimensional quantum system is described by a Hamiltonian operator

$$\hat{H} = -\frac{\hbar^2}{2m}\frac{d^2}{dx^2} + V(|x|)$$

Show that the stationary states are necessarily either even or odd functions.

6 Prove explicitly that $\hat{p}_x \equiv -i\hbar\,\partial/\partial x$ is Hermitian. (Assume $\psi \to 0$ at $x = \pm\infty$.)

7 Use the one-dimensional time-dependent Schrödinger equation to show that for some arbitrary quantum state ψ

$$\frac{d}{dt}\langle \hat{p}_x \rangle = -\left\langle \frac{\partial \hat{V}}{\partial x} \right\rangle$$

and

$$m\frac{d}{dt}\langle \hat{x} \rangle = \langle \hat{p}_x \rangle$$

How do you interpret these equations in the classical limit?

8 Use the commutation relation equation (6.59) to prove

$$x^n\hat{p}_x - \hat{p}_x x^n = ni\hbar x^{n-1}$$

Hence deduce

$$[f(x), \hat{p}_x] = i\hbar\,df/dx$$

9 A particle is described by the classical Hamiltonian

$$H = p^2/2m + k_1 x + k_2 y$$

where k_1 and k_2 are constants. The system is quantized. Evaluate $[\hat{H}, \hat{p}_x]$, $[\hat{H}, \hat{p}_y]$, $[\hat{H}, \hat{p}_z]$ and $[\hat{H}, \hat{x}]$. Which, if any, of $\langle \hat{p}_x \rangle$, $\langle \hat{p}_y \rangle$ and $\langle \hat{p}_x \rangle$ is conserved?

10 If $\hat{A}$ is an Hermitian operator and ψ_n represent energy eigenstates with eigenvalues E_n, show that

$$\langle \psi_n|[\hat{H}, \hat{A}]|\psi_m \rangle = (E_n - E_m)\langle \psi_n|\hat{A}|\psi_m \rangle$$

Hence deduce

$$\langle \psi_n|\hat{p}|\psi_m \rangle = im(E_n - E_m)\langle \psi_n|\hat{x}|\psi_m \rangle/\hbar$$

for a particle of mass m with Hamiltonian as in equation (6.30).

11 Evaluate $\langle \hat{x}^2 \rangle$ and $\langle \hat{p}^2 \rangle$ for the stationary states of the infinite square-well system treated in section 2.2. Use the definition of Δx

and Δp given in section 6.4 to verify that the Heisenberg uncertainty principle, $\Delta x \, \Delta p \geq \hbar/2$, is satisfied for all these states.

12 Show that the ground state of the simple harmonic oscillator actually realizes the strict minimum $\Delta x \, \Delta p = \hbar/2$.

13 Using the results of Exercise 4.7 extend Exercise 6.12 above to cover excited states to show that $\Delta x \, \Delta p = (n + \frac{1}{2})\hbar$. [Hint: remember $\langle \hat{T} \rangle = \langle \hat{p}^2 \rangle/2m = \langle V \rangle = \frac{1}{2}K\langle \hat{x}^2 \rangle$ for the simple harmonic oscillator.]

14 Prove explicitly that if a quantum system is in an eigenstate of an observable α represented by an Hermitian operator $\hat{A}$, then the uncertainty in the measurement of α given by $\Delta \alpha = \sqrt{(\langle \hat{A}^2 \rangle - \langle \hat{A} \rangle^2)}$ is identically zero.

15 Repeat Exercise 4.6 to demonstrate explicitly the orthogonality of the simple harmonic oscillator wave functions:

$$\int_{-\infty}^{\infty} H_n(z)H_m(z)e^{-z^2}\,dz = 0 \qquad n \neq m$$

16 Use the generating functional equation for the simple harmonic oscillator given in Exercise 4.4 to deduce the elements of the (infinite) matrix representation of $\hat{x}$ for this system:

$$x_{mn} = \int_{-\infty}^{\infty} u_m^*(x)xu_n(x)\,dx = \begin{cases} \dfrac{1}{\alpha}\left[\dfrac{n+1}{2}\right]^{1/2} & m = n+1 \\[2ex] \dfrac{1}{\alpha}\left[\dfrac{n}{2}\right]^{1/2} & m = n-1 \\[2ex] 0 & \text{otherwise} \end{cases}$$

By multiplying two such matrices deduce the matrix for $\hat{x}^2$. Check the diagonal elements of your result by comparing with $\langle \hat{x}^2 \rangle$ calculated in Exercise 4.7.

17 Using the closure relation (C.9) given in Appendix C, prove the identity

$$\langle \psi_m | \hat{A}\hat{B} | \psi_k \rangle = \sum_n \langle \psi_m | \hat{A} | \phi_n \rangle \langle \phi_n | \hat{A} | \psi_k \rangle$$

where the ϕ_n form a complete orthonormal set of states.

Chapter 7

1 Consider the matrices for $\hat{L}_x$ and $\hat{L}_y$ in the $l = 1$ case given by equation (7.30). Show that their eigenvalues are 0, $\pm \hbar$. Find the

corresponding eigenvectors, normalize them and verify that they are orthogonal. Repeat the question for $\hat{S}_x$ and $\hat{S}_y$, showing that the eigenvalues are $\pm\hbar/2$.

2 Show that $\int \psi(\hat{L}_+\phi)^*\,d\tau = \int \phi^*(\hat{L}_-\psi)\,d\tau$, where ϕ and ψ are arbitrary wave functions and $\hat{L}_\pm$ are defined by equation (7.33). Hence prove

$$\int |\hat{L}_+ Y_{lm}|^2\,d\tau = \int Y_{lm}^*\hat{L}_-\hat{L}_+ Y_{lm}\,d\tau$$
$$= \hbar^2[l(l+1) - m(m+1)]$$

3 Show for the angular momentum eigenstate $|l, m\rangle$ that $\langle \hat{L}_x \rangle = \langle \hat{L}_y \rangle = 0$. Hence (using the exact definition of Δ) prove the uncertainty relation

$$\Delta L_x \Delta L_y = \tfrac{1}{2}[l(l+1) - m^2]\hbar^2$$
$$\geqslant \tfrac{1}{2}|m|\hbar^2$$

[Hint: by symmetry $\Delta L_x = \Delta L_y$.]

4 Prove that the Pauli spin matrices satisfy the following identities

$$\hat{\sigma}_x^2 = \hat{\sigma}_y^2 = \hat{\sigma}_z^2 = 1$$
$$\hat{\sigma}_x\hat{\sigma}_y + \hat{\sigma}_y\hat{\sigma}_x = \hat{\sigma}_x\hat{\sigma}_z + \hat{\sigma}_z\hat{\sigma}_x = \hat{\sigma}_y\hat{\sigma}_z + \hat{\sigma}_z\hat{\sigma}_y = 0$$

5 Show for the spin state $\begin{bmatrix} a \\ b \end{bmatrix}$ that $\Delta S_z = 0$ if, and only if, $a = 0$ or $b = 0$.

6 A particle is observed initially to have spin component $+\hbar/2$ along the z axis. A measurement of its spin component along another axis z' is then made, where the angle between z and z' is $60°$. What is the probability of obtaining the value $-\hbar/2$?

7 Suppose that in Exercise 7.6 the measurement along z' is not recorded. A subsequent re-measurement of the z component of spin is then made. What is the probability that the spin will now point *down* the z axis, i.e. have eigenvalue $-\hbar/2$?

8 An electron with no orbital angular momentum is subjected to a uniform magnetic field B oriented in the z direction. At time $t = 0$ the electron is in an eigenstate of $\hat{S}_x$ with eigenvalue $+\hbar/2$. Using equation (7.67) show that at time t the state is described by

$$\frac{1}{\sqrt{2}} \begin{bmatrix} e^{-1/2\,i\omega t} \\ e^{1/2\,i\omega t} \end{bmatrix}$$

where $\omega = eB/m_e$. Hence deduce that the state will again be an eigenvector of $\hat{S}_x$ after a time $t = \pi/\omega$ but with eigenvalue $-\hbar/2$. When will the system return to its initial state? Show that at a general time

t the state is an eigenstate of the operator $\hat{S}_\phi \equiv \hat{S}_x \cos\phi + \hat{S}_y \sin\phi$, where $\phi = \omega t$.

Chapter 8

1 A particle is in an orbital angular momentum eigenstate $|l, m\rangle$. The position probability $P(r, t)$ is generally dependent on θ and ϕ. Show, however, using equation (B.12) of Appendix B, that if P is averaged over all m (for fixed l) then the result is independent of θ and ϕ, i.e. spherically symmetric.

2 A particle moves in a spherically symmetric potential

$$
\begin{aligned}
V &= 0 \qquad r < a \\
&= V_0 \qquad r > a
\end{aligned}
$$

Solve for the radial wave function in the $l = 0$ case. Hence show that the energy levels are given by the roots of the equation $\alpha \cot \alpha a = -\beta$ as in the one-dimensional square-well system treated in section 2.4.

3 Repeat Exercise 8.2 for the $l = 1$ case using the solution (8.9) for $r < a$ to show that the energy levels are given by

$$
\frac{\cot \alpha a}{\alpha a} - \frac{1}{\alpha^2 a^2} = \frac{1}{\beta a} + \frac{1}{\beta^2 a^2}
$$

[Hint: the solution for $r > a$ in the $l = 1$ case is of the form $(A/r + B/r^2)e^{-\beta r}$.]

4 The neutron and proton that make up the deuteron may be modelled by considering a single particle of the reduced mass of the system moving in a square-well potential. The binding energy is measured to be 2.21 MeV. If the depth of the well is taken to be 41.6 MeV, calculate the width of the well, i.e. the effective range of the nuclear force, to three significant figures. (You may neglect the small mass difference between the neutron and the proton.)

5 Calculate $\langle \hat{r} \rangle$ for the ground state of the infinite spherically symmetric square-well system.

6 Calculate $\langle \hat{r} \rangle$ and $\langle \hat{V} \rangle$ for a hydrogen atom in the states $|n, l, m\rangle = |2, 1, \pm 1\rangle$.

7 The general form of the radial wave function $R_{nl}(r)$ for the hydrogen atom is, in the special case $l = n - 1$,

$$
[(2n)!]^{-1/2}(2/na_0)^{3/2}(2r/na_0)^{n-1}\exp(-r/na_0)
$$

Find $\langle \hat{r} \rangle$ and $\langle \hat{r}^2 \rangle$. Hence show that

$$\Delta r = \langle \hat{r} \rangle / (2n + 1)^{1/2}$$

How do you interpret this equation classically for large n?

Chapter 9

1 Convince yourself that the four operators specified in equation (9.10) all commute with each other; similarly the four operators specified in equation (9.11).

2 Attempt the algebra omitted between equations (9.15) and (9.16). It is laborious, and in case of failure it can be found in section 17.1 of *Quantum Mechanics* by Amit Goswami, published by Wm. C. Brown in 1992.

3 Enumerate the possible values of the quantum numbers j and m_j for states in which $l = 2$ and $s = \frac{1}{2}$. Repeat for $l = 3$ and $s = \frac{1}{2}$.

4 Discuss the splitting of the $l = 2$ state of the one-electron atom due to (i) spin-orbit coupling, (ii) the strong-field Zeeman effect and (iii) the weak-field Zeeman effect.

5 Discuss spin–orbit coupling and the strong- and weak-field Zeeman effects in the case of an energy level of helium whose total orbital and total spin quantum numbers are 2 and 1 respectively.

Chapter 10

1 Show that a first-order perturbation of the form αx ($\alpha =$ constant) does not alter the spacing of the energy levels of the simple harmonic oscillator.

2 Write down the energy eigenfunctions and eigenvalues of a particle moving between infinite potential barriers placed at $x = \pm a$. Derive the first-order shift in the energy levels resulting from the introduction of a small repulsive potential $V'(x)$ where

$$V'(x) = \begin{cases} V_0 & \text{for } |x| \leqslant b \quad \text{where } b < a \\ 0 & \text{for } a > |x| > b \end{cases}$$

Discuss how the shift in energy levels depends on the parity of the wave function if $b \ll a$.

3 The effect of the finite size of the nucleus in the hydrogen atom can be accounted for by using perturbation theory with the following

additional term in the Hamiltonian:

$$\hat{H}' = \begin{cases} \dfrac{-e^2}{8\pi\varepsilon_0 r_0}\left(3 - \dfrac{r^2}{r_0^2}\right) + \dfrac{e^2}{4\pi\varepsilon_0 r} & \text{for } r \leqslant r_0 \\ 0 & \text{for } r > r_0 \end{cases}$$

where r_0 is the radius of the nucleus which is here considered to be a sphere of uniform charge density.

  (i) By considering what happens to the Coulomb field at points *inside* the nucleus, justify the form of $\hat{H}'$ above.

  (ii) Write down an integral for the first-order shift in energy of the state with $n = 2$ and $l = 0$ (i.e. the 2s state), and one of the states with $n = 2$ and $l = 1$ (i.e. one of the 2p states).

  (iii) Using the approximation $r_0 \ll a_0$ (the Bohr radius), show that the fractional shift in energy of the 2s state is of order $(r_0/a_0)^2$ while for the 2p states it is of order $(r_0/a_0)^4$.

4 Although to a first approximation the vibrations of a diatomic molecule such as CO may be described in terms of a quantum mechanical simple harmonic oscillator, a better approximation is to work with a perturbed Hamiltonian in the form

$$\hat{H} = \frac{\hat{p}^2}{2m} + \frac{kx^2}{2} + ax^3 + bx^4$$

This form includes the small perturbation $\hat{H}' = ax^3 + bx^4$ where a and b are small constants. Show that, to first order, the energy of the perturbed ground state is

$$\tfrac{1}{2}\hbar\omega + \tfrac{3}{4}b(\hbar/m\omega)^2$$

5 Use the results of Exercise 4.7 to calculate the first-order shift in the energy levels of a simple harmonic oscillator subject to a perturbation αx^2 ($\alpha = $ constant). Compare your result with the exact solution obtained by re-scaling the force constant $K \to K + 2\alpha$.

6 A particle moves in one dimension on a circle of circumference L, i.e. in a one-dimensional box $|x| \leqslant L/2$ with periodic boundary conditions. A perturbing potential $-V_0 \exp(-x^2/a^2)$, where $a \ll L$, is introduced. Show that the energy levels are, to first order, split by $\sqrt{\pi}(V_0 a/L)\exp(-a^2k^2)$, where $k = 2n\pi/L$ ($n = 1, 2, 3, \ldots$).

7 Ignoring spin–orbit effects, the first ($n = 2$) excited state of hydrogen is four-fold degenerate: the states $|n, l, m\rangle = |2, 0, 0\rangle$, $|2, 1, -1\rangle$, $|2, 1, 0\rangle$ and $|2, 1, 1\rangle$ all have the same energy. If,

however, an electric field E is applied the degeneracy is partially lifted. Treat the interaction energy $-eEz = -eEr\cos\theta$ as a perturbation and show that the only non-vanishing matrix elements involving the four degenerate states are

$$H'_{12} = H'_{21} = 3eEa_0$$

where 1 and 2 refer to $|2,0,0\rangle$ and $|2,1,0\rangle$ respectively.

Diagonalize the 4×4 H' matrix to show that the field splits the quadruplet into three levels with energy shifts $0, \pm 3eEa_0$. Calculate the magnitude of this splitting for an applied field of $10^6 \, \text{V m}^{-1}$.

8 A particle moves in the oscillator potential λx^4, $\lambda = $ constant, and is in the ground state. Using the (normalized) trial wave function $\alpha^{1/2}\pi^{-1/4}\exp(-\alpha^2 x^2/2)$, where α is an adjustable parameter, use the variational method to estimate the energy of the ground state.

9 Repeat the calculation given at the end of subsection 10.3.1 using instead the (normalized) trial wave function

$$\begin{array}{ll} a^{-1/2}\cos(\pi x/2a) & |x| < a \\ 0 & |x| > a \end{array}$$

where a is an adjustable parameter, to estimate the energy of the ground state of the simple harmonic oscillator.

10 A particle is confined to a one-dimensional box $0 < x < 1$ with impenetrable walls. Compare the variational estimates of the ground state energy using the two (unnormalized) trial wave functions $x(1-x)$ and $x(1-x) + x^2(1-x)^2$.

11 Estimate the ground state energy of the hydrogen atom using the variational method, taking the unnormalized trial function $(1 - r/\lambda)$ for $r \leqslant \lambda$, and zero for $r > \lambda$.

Does your result for λ_{min} also give a reasonable estimate of the size of the hydrogen atom, characterized by the Bohr radius $a_0 = 4\pi\varepsilon_0\hbar^2/me^2$?

12 Determine approximately the deuteron ground state energy for the potential $V(r) = -A\exp(-r/a)$ (with $A = 32 \, \text{MeV}$, $a = 2.2 \times 10^{-15} \, \text{m}$) using the variational principle. Use as trial wave function a function of the form $R = c\exp(-\alpha r/2a)$, depending on one parameter α. The value of the constant c follows from the normalization condition

$$\int_0^\infty R^2 r^2 \, dr = 1$$

Chapter 11

1 At time $t = 0$ a hydrogen atom in its ground state is subject to the electric field $E_0 e^{-t/\tau}$ ($E_0 = $ constant). Find the first-order probability that the atom will be found in the excited state $|2,1,0\rangle$ at $t \to \infty$.

2 Some short-lived subnuclear particles of roughly a proton mass have half-lives $\approx 10^{-24}$ s. Show that there is a high fractional uncertainty in the masses of such particles.

3 Consider a particle confined to an infinite one-dimensional square well in an initial stationary state $u_n(x)$. Show that if the particle is subjected to the perturbation $V_0 x \cos \omega t$ it can only make a transition to a final state $u_m(x)$ if $m + n$ is odd.

4 (a) Suppose the stationary states of an electron have the angular dependence $Y_{lm}(\theta, \phi)$. Show that the matrix element $\langle m|z|m'\rangle$ vanishes unless $\Delta m \equiv m - m' = 0$. Hence deduce that the emission of plane-polarized electric dipole radiation by an electron in a spherically symmetric potential is forbidden between states of different m quantum number. (b) Similarly, show that circularly polarized electric dipole radiation described by the Cartesian component combinations $x \pm iy$ is forbidden unless $\Delta m = \pm 1$.

5 Using the wave functions given in Table 8.1, compare the relative transition rates for a hydrogen atom in the state $|3,1,0\rangle$ to decay by spontaneous radiative emission to either the state $|2,0,0\rangle$ or the ground state $|1,0,0\rangle$. [Hint: recall the selection rules in Exercise 11.4.]

APPENDIX G

Answers to exercises

Chapter 1
2   5×10^{30}
4   About 10^{27} s.

Chapter 2
1   $\frac{1}{3}a^2[1 - 6/n^2\pi^2]$.

Chapter 4
1   0.314.
7   $(n + \frac{1}{2})\hbar/m\omega$.
9   3.78×10^{-4}; 0.238.
14 7.69×10^{-11} m;
    $1/9$; 0.135; 0.750.

Chapter 5
3   $(4\pi/k^2)\sin^2\delta_0$.

Chapter 6
2   $5\hbar^2/ma^2$; 0.999.
3   $2\pi^2\hbar^2/3ma^2$; 0.961; 0.

Chapter 7
6   $\frac{1}{4}$.
7   $\frac{3}{4}$.
8   $4\pi/\omega, 8\pi/\omega, 12\pi/\omega\ldots$.

Chapter 8
4   1.74×10^{-15} m.
5   $a/2$.
6   $6a_0, -e^2/16\pi\varepsilon_0 a_0$.

Chapter 10

2   $\dfrac{V_0}{a}\left[b \pm \dfrac{a}{n\pi}\sin\dfrac{n\pi b}{a}\right]$

5   $\hbar\alpha/2m\omega$.
7   2.54×10^{-4} eV.
8   $1.08\,(\hbar^4\lambda/4m^2)^{1/3}$.
9   $0.568\hbar\omega$.

11 $E_0 = -\dfrac{5\hbar^2}{16ma_0^2}$

    $\lambda_{\min} = 4a_0$

12 $-2.2\,\text{MeV}$

    $= -3.5 \times 10^{-13}\,\text{J}$.

Chapter 11

1   $\dfrac{2^{15}e^2E_0^2a_0^2}{3^{10}\hbar^2}\left(\dfrac{\tau^2}{1 + \omega^2\tau^2}\right)$

    where $\omega = (E_2 - E_1)/\hbar$.
5   $(8/5)^{12}/8$

Index